99 Points of Intersection

Examples—Pictures—Proofs

Current Printing (last digit):
10 9 8 7 6 5 4 3 2 1

99 Points of Intersection

Examples—Pictures—Proofs

by

Hans Walser

Translated from the original German by
Peter Hilton and Jean Pedersen

Published and Distributed by
The Mathematical Association of America

SPECTRUM SERIES

The Spectrum Series of the Mathematical Association of America was so named to reflect its purpose: to publish a broad range of books including biographies, accessible expositions of old or new mathematical ideas, reprints and revisions of excellent out-of-print books, popular works, and other monographs of high interest that will appeal to a broad range of readers, including students and teachers of mathematics, mathematical amateurs, and researchers.

777 Mathematical Conversation Starters, by John de Pillis

99 Points of Intersection: Examples—Pictures—Proofs, by Hans Walser. Translated from the original German by Peter Hilton and Jean Pedersen

All the Math That's Fit to Print, by Keith Devlin

Carl Friedrich Gauss: Titan of Science, by G. Waldo Dunnington, with additional material by Jeremy Gray and Fritz-Egbert Dohse

The Changing Space of Geometry, edited by Chris Pritchard

Circles: A Mathematical View, by Dan Pedoe

Complex Numbers and Geometry, by Liang-shin Hahn

Cryptology, by Albrecht Beutelspacher

Five Hundred Mathematical Challenges, Edward J. Barbeau, Murray S. Klamkin, and William O. J. Moser

From Zero to Infinity, by Constance Reid

The Golden Section, by Hans Walser. Translated from the original German by Peter Hilton, with the assistance of Jean Pedersen.

I Want to Be a Mathematician, by Paul R. Halmos

Journey into Geometries, by Marta Sved

JULIA: a life in mathematics, by Constance Reid

R. L. Moore: Mathematician and Teacher, by John Parker

The Lighter Side of Mathematics: Proceedings of the Eugène Strens Memorial Conference on Recreational Mathematics & Its History, edited by Richard K. Guy and Robert E. Woodrow

Lure of the Integers, by Joe Roberts

Magic Tricks, Card Shuffling, and Dynamic Computer Memories: The Mathematics of the Perfect Shuffle, by S. Brent Morris

Martin Gardner's Mathematical Games: The entire collection of his Scientific American columns

The Math Chat Book, by Frank Morgan

Mathematical Adventures for Students and Amateurs, edited by David Hayes and Tatiana Shubin. With the assistance of Gerald L. Alexanderson and Peter Ross

Mathematical Apocrypha, by Steven G. Krantz

Mathematical Apocrypha Redux, by Steven G. Krantz

Mathematical Carnival, by Martin Gardner

Mathematical Circles Vol I: In Mathematical Circles Quadrants I, II, III, IV, by Howard W. Eves

MAA Service Center
P.O. Box 91112
Washington, DC 20090-1112
800-331-1622   FAX 301-206-9789

Author's Foreword

The 99 points of intersection presented here were collected during a year-long search for surprising concurrence of lines. For each example we find compelling evidence for the sometimes startling fact that in a geometric figure three straight lines, or sometimes circles, pass through one and the same point. Of course, we are familiar with some examples of this from basic elementary geometry — the intersection of medians, altitudes, angle bisectors, and perpendicular bisectors of sides of a triangle. Here there are many more examples — some for figures other than triangles, some where even more than three straight lines pass through a common point.

I have been asked to present the points of intersection purely visually, without caption and verbal commentary. This reading — and picture — book should encourage the reading and understanding of pictures. It should also encourage readers to find such points of intersection for themselves. Extremely helpful for discovering and testing points of intersection are interactive Geometry-software programs, such as *Cabri Geometry II*, *Geometer's Sketchpad*, *Cinderella*, or *Euklid*, just to name a few. A search for "dynamic geometry software" on the web will reveal these and many other programs that are available in English.

This collection is, of course, in no way complete. Further examples may be found in [Donath 1976] and in the expressly history-of-geometry presentation of [Baptist 1992].

Examples: The first part of this book contains a few general thoughts and examples on the theme of points of intersection.

Pictures: The second — and main — part presents 99 points of intersection in sequence, each preceded by three figures.

Proofs: The third part contains an overview of some typical methods of proving the existence of such points of intersection, and classical theorems on points of intersection, as well as exemplary proofs of a few of the points of intersection previously introduced.

Many examples were brought to me by colleagues. I particularly want to thank Heiner Bubeck of Weingarten, Wolfgang Kroll of Marburg, and Roland Wyss of Flumenthal for a few harder examples. Most examples, however, arose from questions and suggestions made by my students and also by my preservice teachers. To all of these I owe a great debt of gratitude.

For very helpful pedagogical and technical advice in the area of dynamic geometry software I must thank Hans-Jürgen Elschenbroich of Medienzentrum Rheinland, and Heinz Schumann of Weingarten.

My thanks go finally to Herr Jürgen Weiss of the Wissenschaftsverlag "Edition am Gutenbergplatz Leipzig" for helpfully entrusting this new printing to the collection of popular science of Leipzig "Eagle-Einblicke."

Hans Walser
Frauenfeld
July 2004

Foreword to the English Edition

We have been faithful to the original German text, except that we have added some references to publications and web sites in English — without, of course, omitting any of the author's own chosen references to works which are mostly in German. Thus we have, in fact, rendered this English edition as appropriate as possible to English speakers who are not German speakers, without in any way putting those German speakers who might wish to learn the material from an English translation at a disadvantage.

Two remarks related to the content are appropriate here. First, the text reflects a particular interest of the author. In this respect it differs from the content of the two previous translations of work by the author in this series (*Symmetry* and *The Golden Section*). In this book the author has given expression to his interest in an aspect of geometry that has not previously been given individual treatment. Many of the examples of points of intersection are well known but, naturally, the organization of the material is highly innovative. Moreover, the author has felt free to draw on his wide knowledge of geometric method in devising appropriate proofs.

Second, the author has also exploited his experience in the uses of modern technology in devising and presenting mathematical arguments. This translation has been supplemented by references to English language sources which will help readers appreciate these arguments even if they were not already familiar with them.

We have (as in our previous translations) made a few modifications of the original text in order that the English version should not read too obviously like a translation from German — or any other foreign language!

It is again a pleasure to express our appreciation to our colleague Jerry Alexanderson for his careful and critical perusal of this translation.

PETER HILTON   JEAN PEDERSEN
Binghamton   Santa Clara
                August, 2005

Author's Note to the English Edition

It is a great pleasure to express once more my appreciation of the careful work done by my colleagues Peter Hilton and Jean Pedersen in making available an English version of my text.

I would also like to take this opportunity again to thank Jerry Alexanderson for his editorial work, and Beverly Ruedi and Elaine Pedreira for their careful attention to detail in the production of this work.

HANS WALSER
Frauenfeld
August 2005

Contents

CHAPTER 1

What's it all about?

1.1 If three lines meet

Three or more arbitrary straight lines or curves generally have no common point of intersection. But if such a common point of intersection exists, there comes the inevitable question — Why? Why, on a photograph, do the extensions of the space-edges pass through a common point, called the vanishing point? This is obviously the case if the lines are in fact parallel, thus having a very special relative position.

1.1.1 The dodecagon

Very often the existence of a point of intersection of three or more straight lines can be explained by considerations of symmetry. Thus the six midpoint-diagonals of a regular 12-gon trivially pass through a common point, the midpoint of the 12-gon (Figure 1).

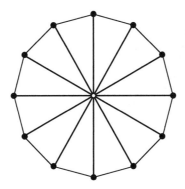

FIGURE 1
Six diagonals through the midpoint

The three diagonals of Figure 2 plainly intersect too, on grounds of symmetry, since the two oblique diagonals are situated as mirror images in the vertical diagonal.

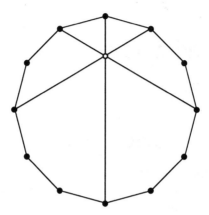

FIGURE 2
A common point of intersection on grounds of symmetry

Things are different in the example of Figure 3. Do the three straight lines actually pass through one and the same point?

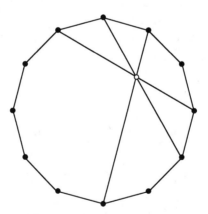

FIGURE 3
Do the 3 diagonals intersect in a point?

If the three diagonals meet in a point, then, on grounds of symmetry, there must be a fourth diagonal also passing through this point (Figure 4).

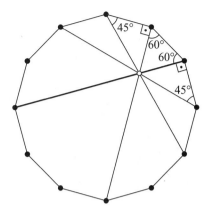

FIGURE 4
The fourth diagonal and special angles

From these diagonals arise, at four adjacent vertices of the dodecagon, angles of 45°, 90°, 60°, 60°, 90°, 45° (Figure 5).

There is a sequence of three triangles, two of which are right-angled isosceles with the side-lengths of the dodecagon as lengths of the shorter sides, the middle triangle being equilateral with the side-length that of the dodecagon. So the three triangles have a vertex in common, and the four diagonals all pass through this vertex.

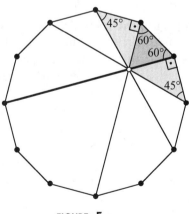

FIGURE 5
Special triangles

1.1.2 A puzzle

This proof-figure is part of a subdivision-puzzle for the regular dodecagon. There emerge eight right-angled isosceles triangles, four small and four large equilateral triangles and a square (Figure 6).

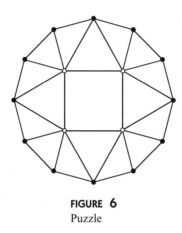

FIGURE 6
Puzzle

1.1.3 Points of intersection of circles

We turn back to Figure 3 and draw, for each of the three diagonals, a circle passing through the center of the dodecagon and having the diagonal as chord (Figure 7). These three circles all, naturally, pass through the midpoint of the dodecagon, but plainly they also intersect in a point outside the dodecagon.

To understand this property we apply the so-called circle-reflection (see [Walser 2000]). These three circles are the images of the three diagonals under reflection in the circumcircle of the dodecagon. The external point of intersection of the three circles is the reflection of the point of intersection of the three diagonals.

We turn back one last time to Figure 3 and draw, for each of the three diagonals, a circle which cuts the circumcircle of the dodecagon orthogonally and has the diagonal as chord (Figure 8).

These three circles clearly have two common points of intersection. To see this, we apply the so-called non-euclidean or hyperbolic geometry (see, [Buchmann 1975], [Cederberg 1995], [Coxeter, 1963], [Filler 1993], [Hartshorne 2000], [Holme 2002], [Kinsey/Moore 2002], [Lenz 1967], [Nöbeling 1976], [Zeitler 1970]). Precisely, we can interpret the circumcircle of the dodecagon with the diagonals as chords as the Klein model of this geometry, and the same circle with the arcs orthogonal to it in its interior as the Poincaré model.

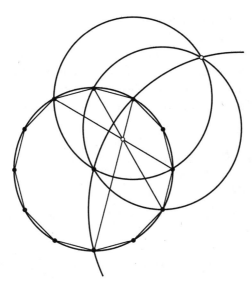

FIGURE 7
The second point of intersection of the circles

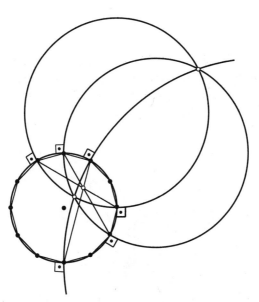

FIGURE 8
The circles orthogonal to the circumcircle

From the point of intersection of the three chords arises the interior point of intersection of the three circles, and conversely. For the external point of intersection we again need to use circle-reflection. The orthogonal circles are invariant under reflection in the circumcircle of the dodecagon, and so are their own reflections. Thus the external point of intersection of the three circles is the reflection of the internal point of intersection.

1.2 Flowers for Fourier

1.2.1 An example

Figure 9 shows three flowers with 5, 7, and 11 petals.

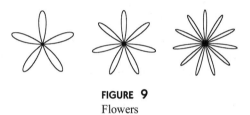

FIGURE 9
Flowers

If we superimpose these three pictures, they clearly have the center and topmost point in common. Do they have other points in common?

The superimposition shows that there are two further points common to all three pictures (Figure 10). Thus, in total, there are four points of intersection of these three pictures.

FIGURE 10
Four points of intersection

FIGURE **11**

The points of intersection lie on a circle and three of them form an equilateral triangle.

These four points lie on a circle; three of them are vertices of an equilateral triangle (Figure 11).

1.2.2 Background

Behind the flowers lurk the functions $y = \cos 5t$, $y = \cos 7t$, $y = \cos 11t$. These are functions of the form $y = \cos nt$; such functions are applied in Fourier-expansions (Jean Baptiste Joseph Fourier, 1768–1830, see [Burg/Haf/Ville 1994], [Butz 2003], [Heuser 1992], [Jänich 2001, p. 266]). Figure 12 shows the graphs of these three functions for the interval $[-\pi, \pi]$ in the usual cartesian representation.

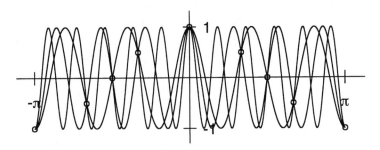

FIGURE **12**

Graphs of the functions

These three function-curves have, for $t \in [-\pi, \pi]$, a total of nine points in common. Now we show the same thing in polar coordinates. For example, we choose for the polar angle φ the polar distance $r(\varphi) = \cos 5\varphi$. In this way a flower appears which, in view of the symmetries of the function $y = \cos 5t$, is plotted twice (Figure 13). Thus we may restrict ourselves to the interval $\varphi \in [0, \pi]$.

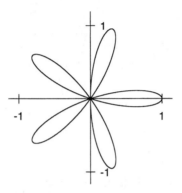

FIGURE 13
Polar representation of $y = \cos 5t$

On esthetic grounds this flower should be turned through 90°, so that one petal points vertically upwards. Our three functions have five common points such that, in the polar representation, the first and last coincide (Why?). Thus, in fact, the flowers have four common points.

Now comes the question: Why were the precise numbers $n = 5, 7, 11$ chosen? Certainly they are three consecutive prime numbers, but that plays no role in our example. A systematic study of the functions $y = \cos nt$ shows that for odd numbers of the form $n = 3k \pm 1$, $k \in \mathbb{Z}$, these functions in the cartesian representation pass through the points indicated in Figure 12. In the polar representation this also holds for even numbers of the form $n = 3k \pm 1$, $k \in \mathbb{Z}$.

Figure 14 shows an example in detail. In this example there are more points through which three or more curves pass, but there is no further point through which all of the curves pass.

Figure 15 shows the situation for $n = 4, 6, 9$; these are numbers of the form $n = 5k \pm 1$, $k \in \mathbb{Z}$. There are six common points of intersection, five of which are vertices of a regular pentagon.

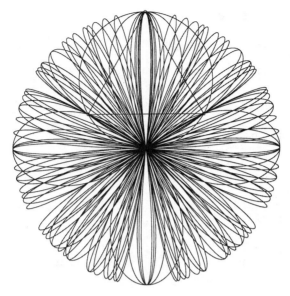

FIGURE 14

$n = 3k \pm 1, k \in \mathbb{Z}$

FIGURE 15

$n = 4, 6, 9$

1.3 Chebyshev and the Spirits

1.3.1 Chebyshev Polynomials

From the Euler identity $e^{i\varphi} = \cos\varphi + i\sin\varphi$ follows, on the one hand,

$$e^{in\varphi} = \cos n\varphi + i\sin n\varphi$$

and, on the other hand,

$$e^{in\varphi} = (\cos\varphi + i\sin\varphi)^n = \sum_{k=0}^{n} \binom{n}{k} \cos^{n-k}\varphi\, i^k \sin^k \varphi.$$

Comparing the real parts yields:

$$\cos n\varphi = \sum_{j=0}^{\lfloor n/2 \rfloor} (-1)^j \binom{n}{2j} \cos^{n-2j}\varphi \sin^{2j}\varphi.$$

Since $\sin^{2j}\varphi = \left(1 - \cos^2\varphi\right)^j$, we can thereby write $\cos n\varphi$ as a polynomial of degree n in $\cos\varphi$. These polynomials are called Chebyshev polynomials (Pafnuti Lwowich Chebyshev, 1821–1894, see [Meyberg/Vachenauer 2001, p. 144], [Madelung 1964, p. 108]). With $x = \cos\varphi$, we employ for $x \in [-1, 1]$ the notation $T_n(x) = \cos(n\arccos x)$. In polynomial notation we have

$$T_0(x) = 1$$
$$T_1(x) = x$$
$$T_2(x) = 2x^2 - 1$$
$$T_3(x) = 4x^3 - 3x$$
$$T_4(x) = 8x^4 - 8x^2 + 1$$
$$T_5(x) = 16x^5 - 20x^3 + 5x$$
$$T_6(x) = 32x^6 - 48x^4 + 18x^2 - 1.$$

The functions are alternately even and odd; they satisfy the recurrence relation

$$T_{n+1}(x) = 2x\,T_n(x) - T_{n-1}(x).$$

The functions oscillate in value over the interval $[-1, 1]$ between -1 and $+1$. Figure 16 shows the graphs of $T_0, \ldots, T_6$.

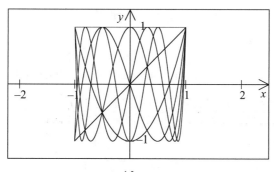

FIGURE 16
Chebyshev polynomials

These graphs can also represent Lissajous curves. The graph of $T_n(x)$ has the parametric representation

$$\overline{x}(t) = \begin{bmatrix} \cos t \\ \cos nt \end{bmatrix}, \quad t \in [0, 2\pi].$$

We recognize from this that the Chebyshev polynomials are closely related to the Fourier functions $y = \cos nt$. It is just a matter of changing the scale on the horizontal axis.

1.3.2 Points of intersection in the Golden Section

We consider now points of intersection of graphs of various Chebyshev polynomials. The graphs of T_2, T_7, and T_{12} (Figure 17) have three points of intersection in common. To describe these points of intersection we use the

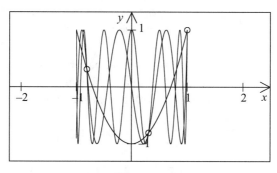

FIGURE 17
Graphs of T_2, T_7, and T_{12}

notation of the Golden Section (see [Walser, 2001, 2004]). We define

$$\tau = \frac{1 + \sqrt{5}}{2} \approx 1.618 \quad \text{and} \quad \rho = \frac{-1 + \sqrt{5}}{2} \approx 0.618.$$

Then we obtain as coordinates of the three points of intersection

$$(1, 1), \ \left(\frac{\rho}{2}, -\frac{\tau}{2}\right), \ \left(-\frac{\tau}{2}, \frac{\rho}{2}\right).$$

1.3.3 An optical effect

Figure 18 shows the graphs of T_0 up to T_{30}.

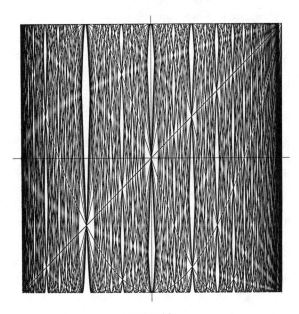

FIGURE 18
Optical effect

We see in this figure some ghostly curves, on which very many points of intersection lie. One of these curves is a parabola with horizontal axis, another is the Lissajous curve of Figure 19 with parametric representation

$$\overline{x}(t) = \begin{bmatrix} \cos t \\ \cos \frac{3}{2}t \end{bmatrix}; \quad t \in [0, 2\pi].$$

The ghostly curves arise from the very many points on them in which several "proper" curves intersect. Thus black is strongly concentrated there and nearby there is much white, thereby making these curves visible.

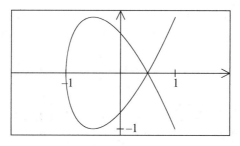

FIGURE 19
The Lissajous curve

1.4 Sheaves generate curves

For where two or three are gathered together,
in my name, there am I in the midst of them.
Matthew 18:20

1.4.1 Sheaves of straight lines

We start with a sheaf of straight lines as in Figure 20.

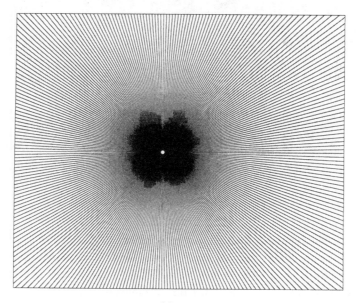

FIGURE 20
A sheaf of straight lines

On this sheaf of straight lines we superimpose a second, congruent sheaf which, however, is laterally displaced (Figure 21). This is best achieved by laying two transparencies, printed with the sheaves, one on top of the other. Then the lateral displacement can be varied to create very interesting effects. One transparency can be slowly twisted. What do we then see?

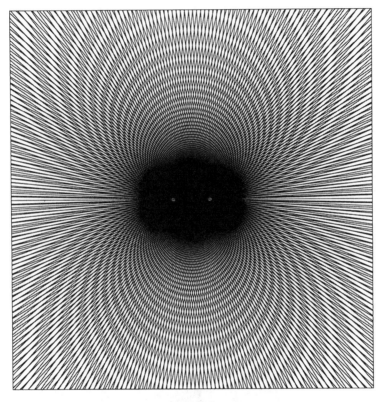

FIGURE 21
Superimposition

We see ghost circles, which actually pass through both sheaf-centers. Really we only see in this picture points of intersection of two straight lines. But these obviously lie on circles which are visible through the white background. Thus we now have points of intersection of three objects.

With these circles arcs are formed where the straight lines cross at angles of the same size.

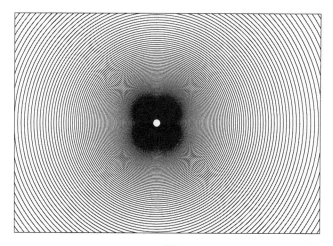

FIGURE 22
Sheaves of circles

1.4.2 Sheaves of circles

We superimpose two congruent sheaves of circles as in Figure 22. The radii of the sheaves of circles are increasing exponentially, giving a better optical effect.

In the superimposition we see additional circles, which in this case are the proportionate circles of Apollonius (Figure 23).

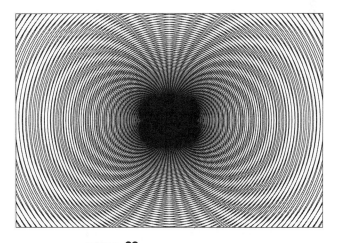

FIGURE 23
Proportionate circles of Apollonius

CHAPTER 2

The 99 points of intersection

The sequences of pictures of the 99 points of intersection are conceived, in the sense of "minimal art," as pictures without words. To this end the following graphic system is employed:

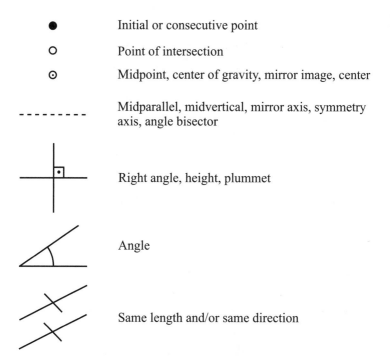

●	Initial or consecutive point
○	Point of intersection
⊙	Midpoint, center of gravity, mirror image, center
- - - - - - - -	Midparallel, midvertical, mirror axis, symmetry axis, angle bisector
	Right angle, height, plummet
	Angle
	Same length and/or same direction

Sometimes there are to be found under the figures references to the literature or hints as to the ideas which gave rise to these figures. (Translator's note: Not all figures have captions.)

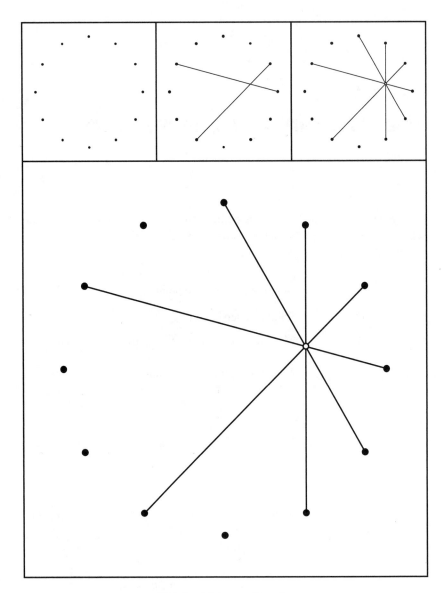

Point of intersection 1
Dodecagon with diagonals

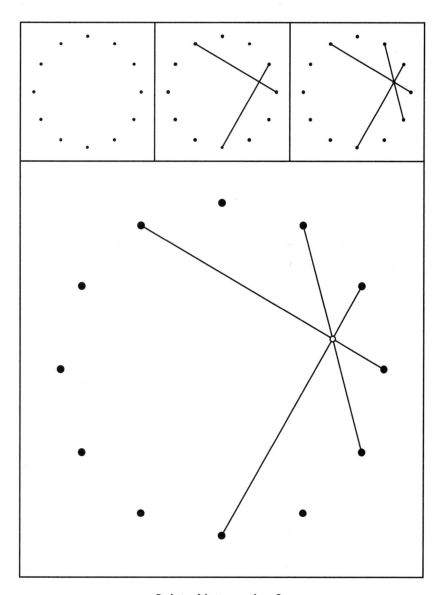

Point of intersection 2
Dodecagon with diagonals

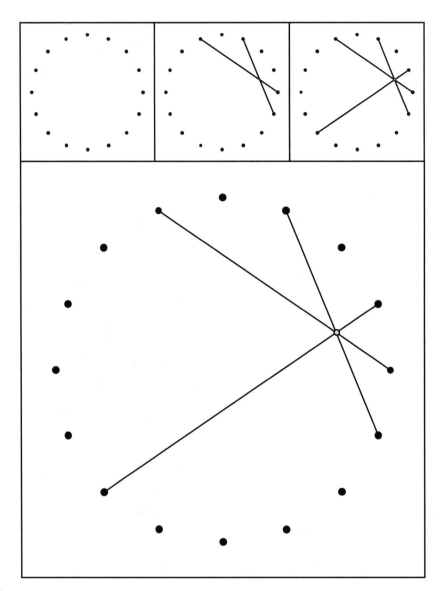

Point of intersection 3
16-gon with diagonals

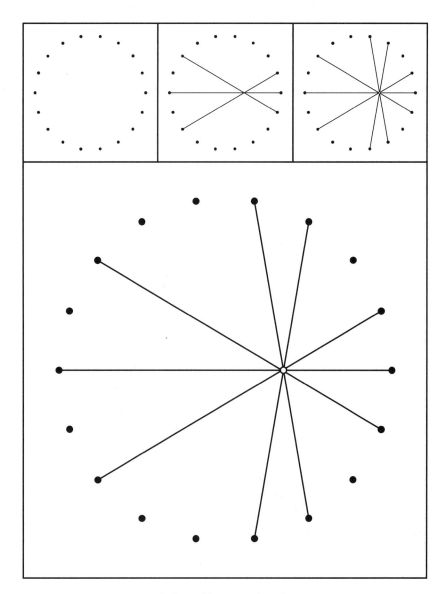

Point of intersection 4
18-gon with diagonals

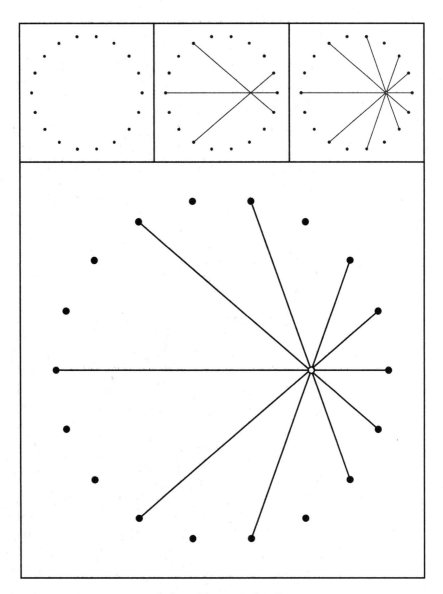

Point of intersection 5
18-gon with diagonals

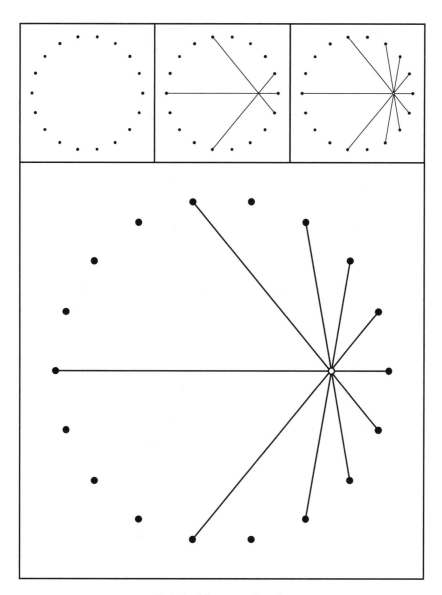

Point of intersection 6
18-gon with diagonals

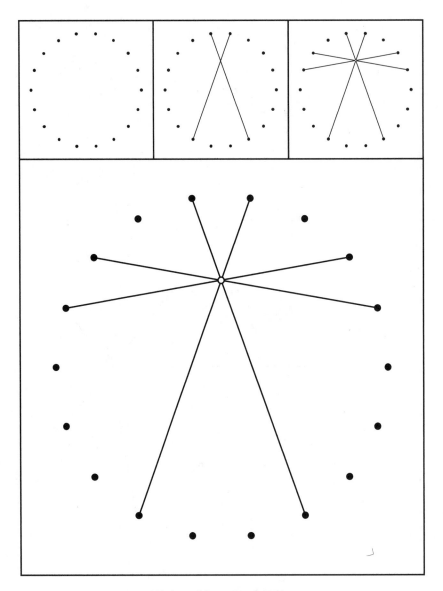

Point of intersection 7

18-gon with diagonals

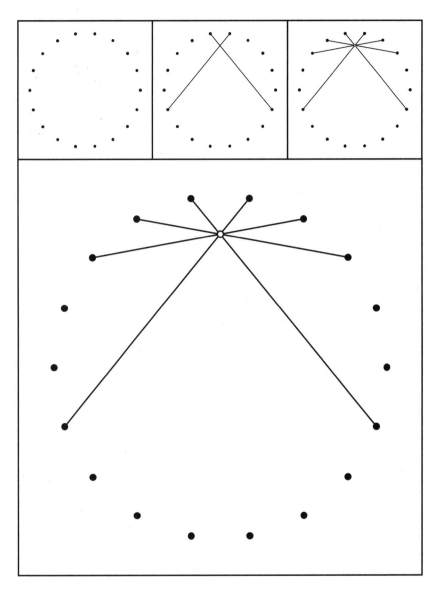

Point of intersection 8
18-gon with diagonals

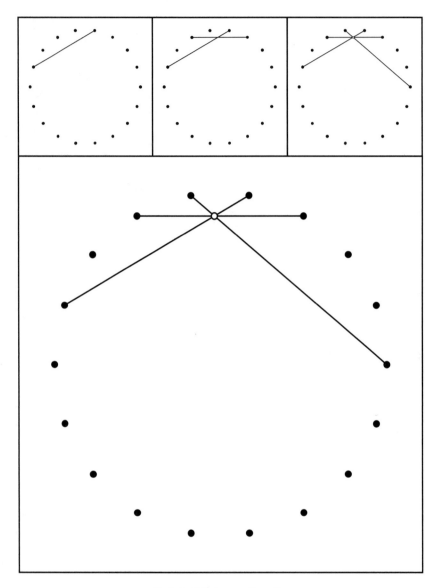

Point of intersection 9
18-gon with diagonals

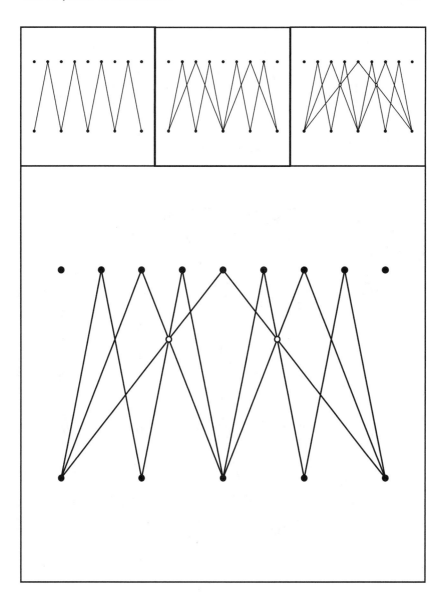

Point of intersection 10
Zigzag

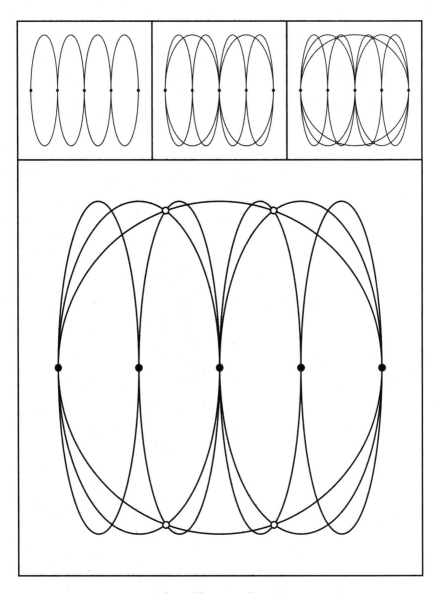

Point of intersection 11
Ellipses

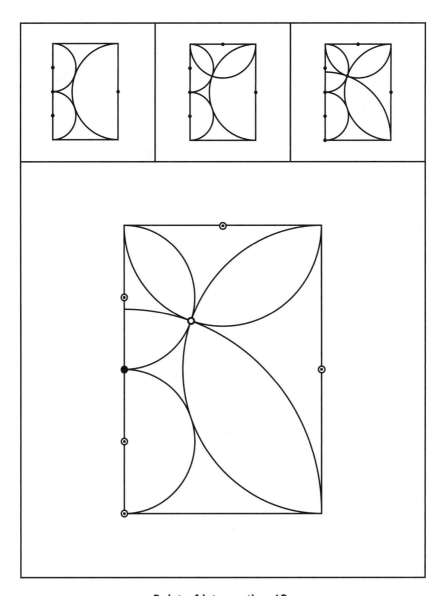

Point of intersection 12
DIN – format (see [Walser 2001, 2004])

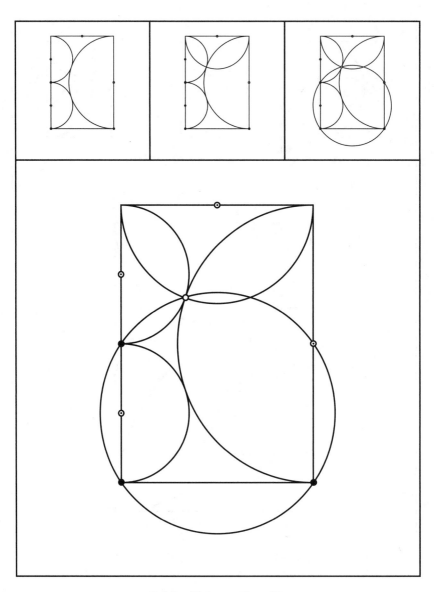

Point of intersection 13
DIN – format

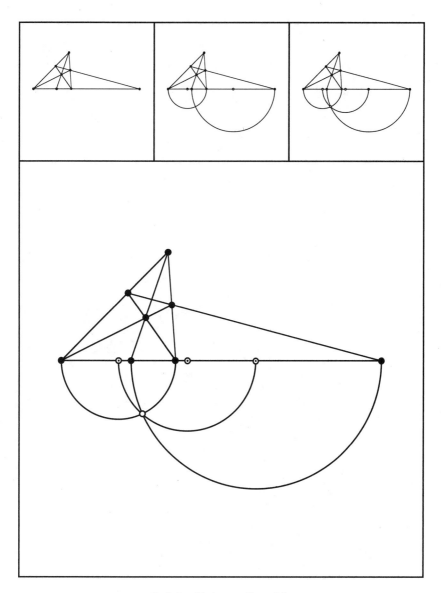

Point of intersection 14
Three circles of Thales

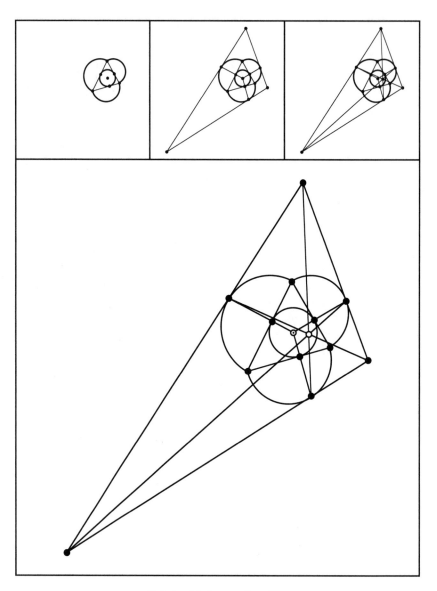

Point of intersection 15
Idea: Wolfgang Kroll

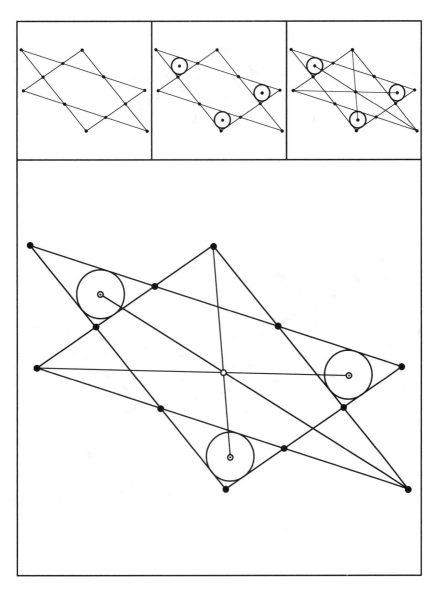

Point of intersection 16
Affinely distorted Star of David with incircles

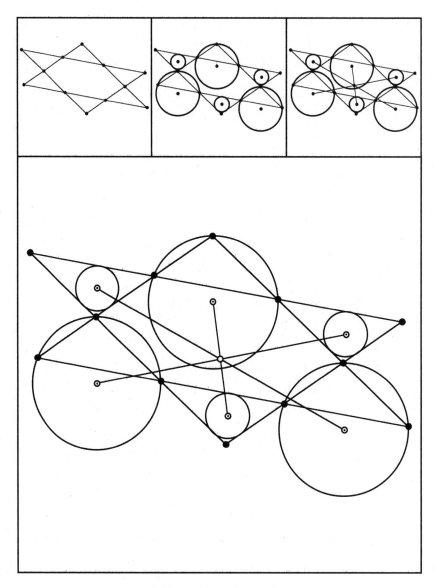

Point of intersection 17
Affinely distorted Star of David with incircles and circumcircles

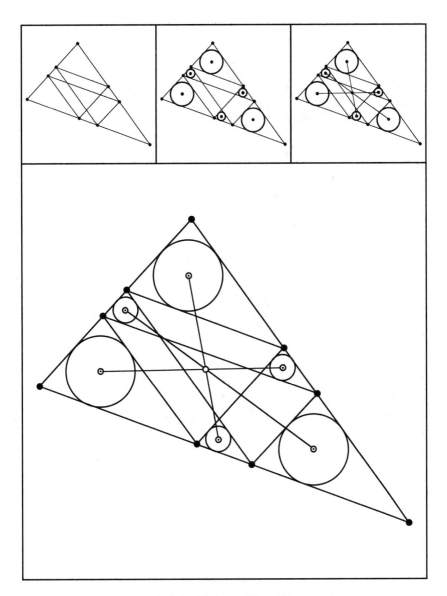

Point of intersection 18
Closed path (see [Kroll 1990]) in a triangle with incircles

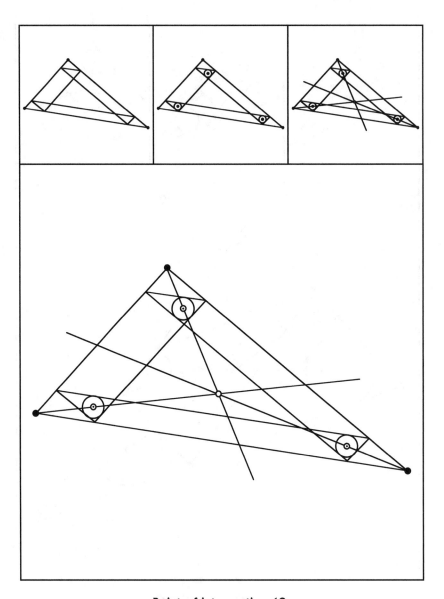

Point of intersection 19

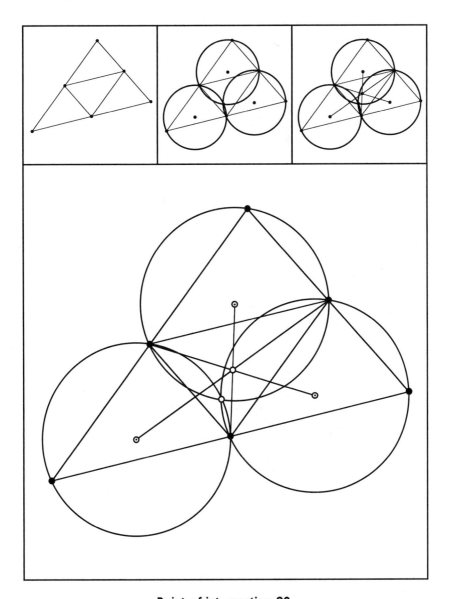

Point of intersection 20
Quartertriangles and circumcircles

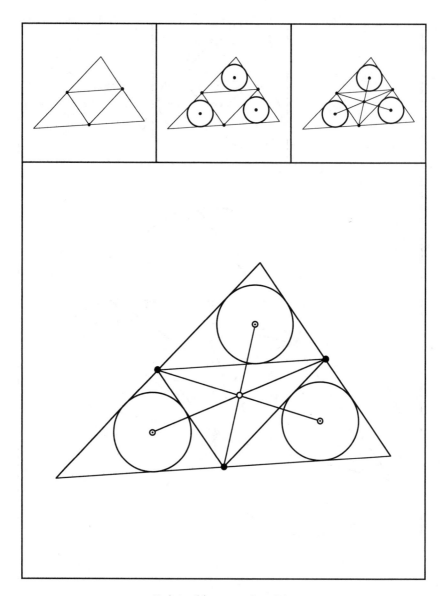

Point of intersection 21
Quartertriangles and incircles

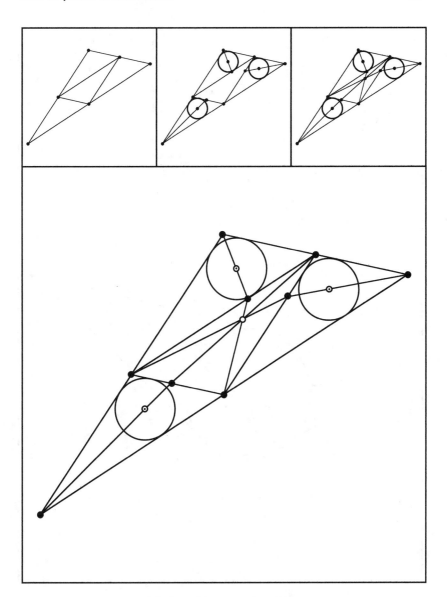

Point of intersection 22
Quartertriangles and incircles

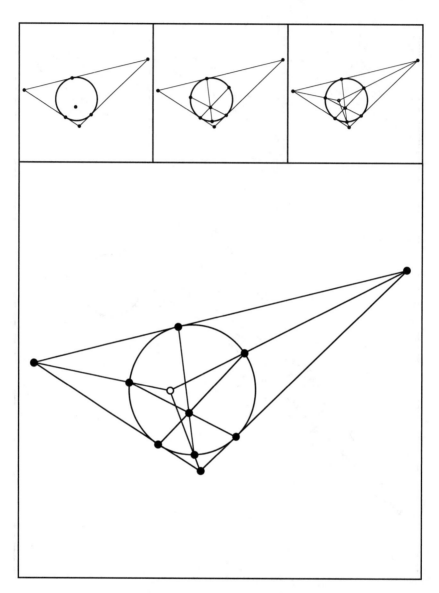

Point of intersection 23
Idea: Wolfgang Kroll

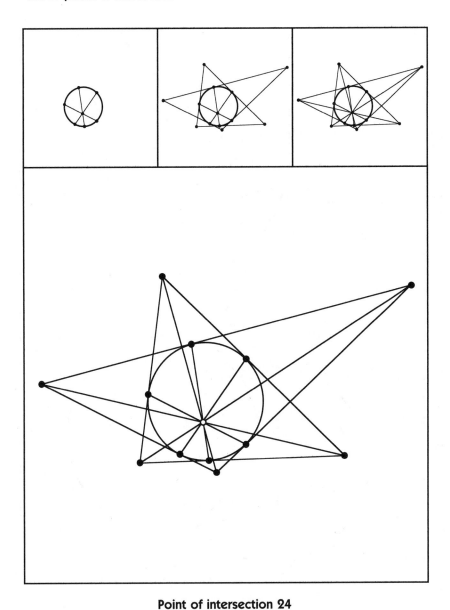

Point of intersection 24
Two triangles with the same incircle. Idea: Wolfgang Kroll

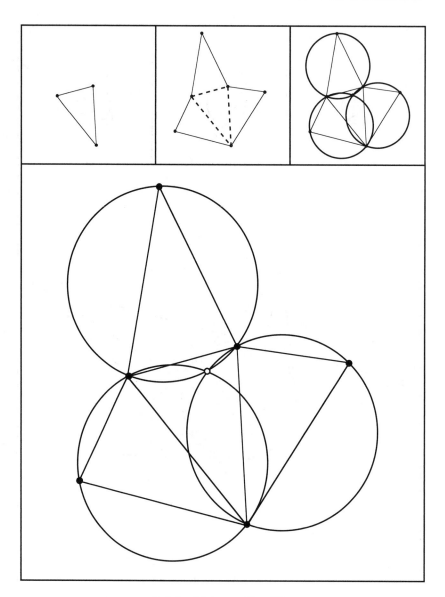

Point of intersection 25
Mirror-image triangles with circumcircles

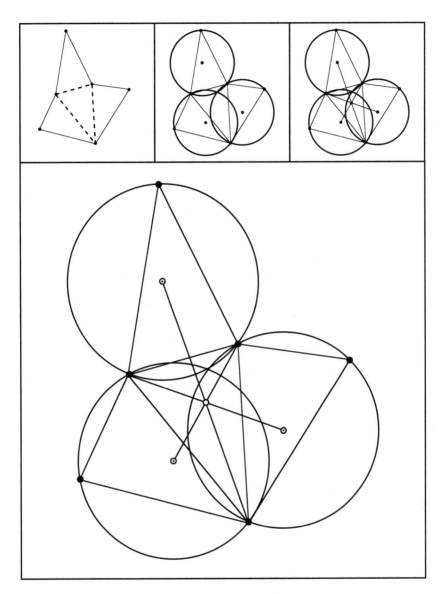

Point of intersection 26

Mirror-image triangles with circumcircles

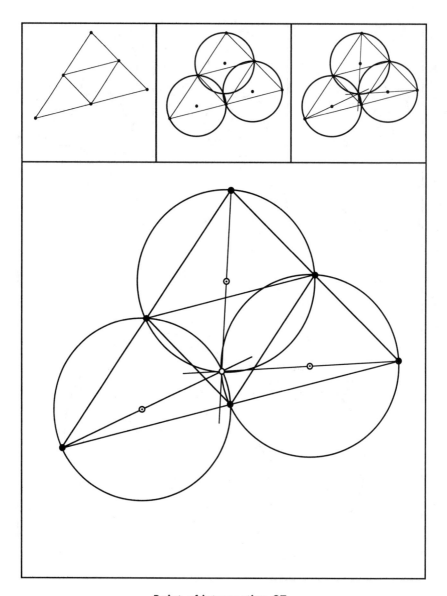

Point of intersection 27

Quartertriangles and circumcircles

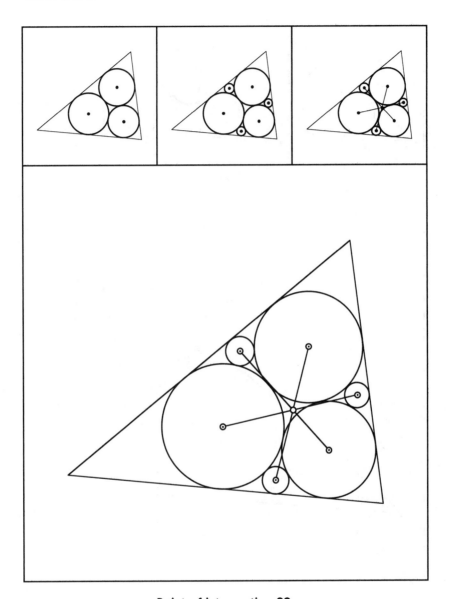

Point of intersection 28
Homage to Gian Francesco Malfatti

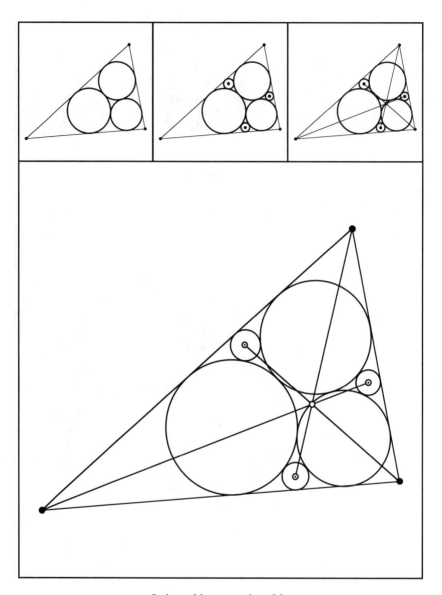

Point of intersection 29

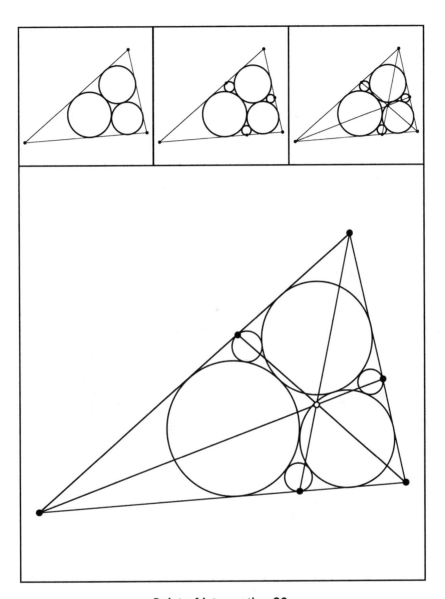

Point of intersection 30

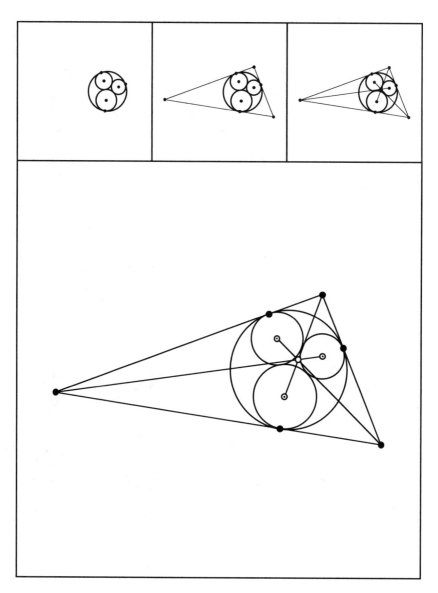

Point of intersection 31

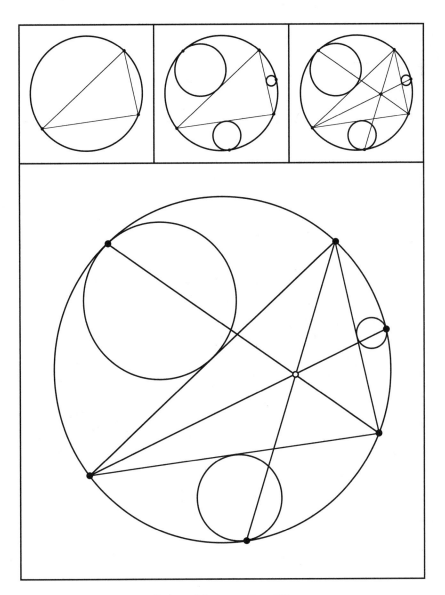

Point of intersection 32

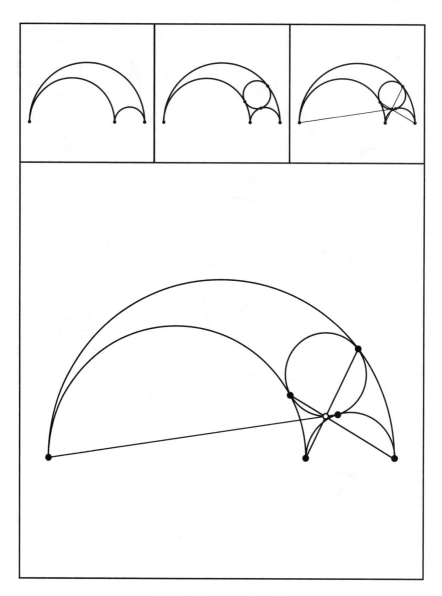

Point of intersection 33
Crescent (Arbelos)

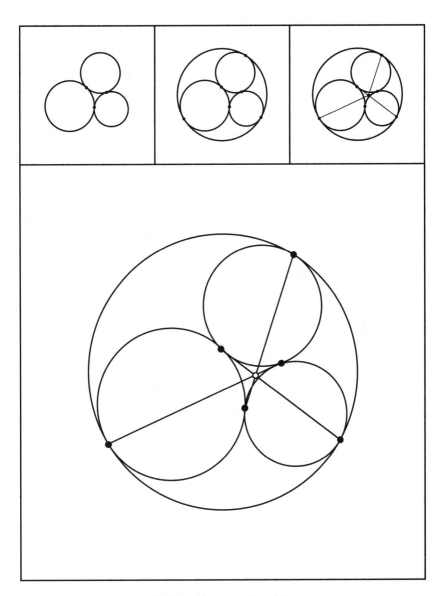

Point of intersection 34
Cloverleaf

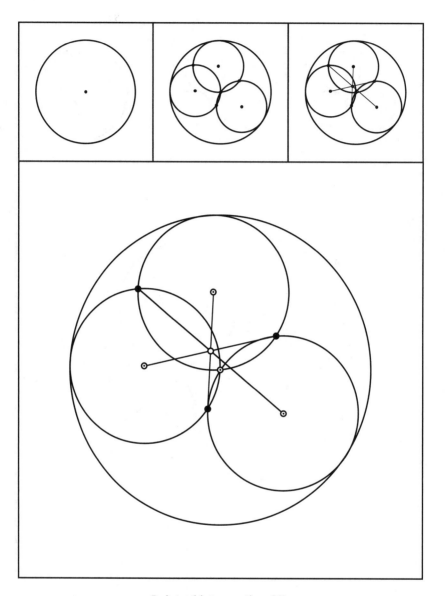

Point of intersection 35

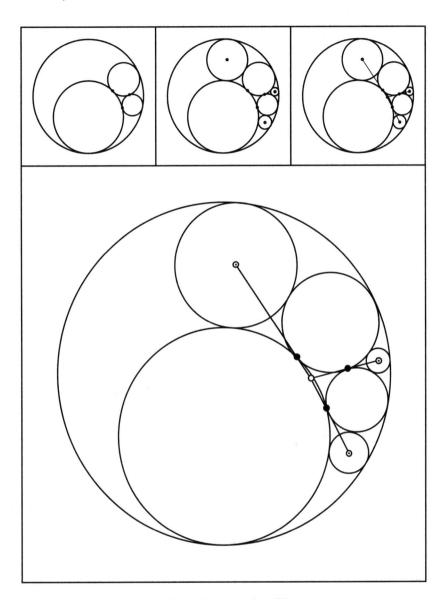

Point of intersection 36
See Remarks 3.6.2

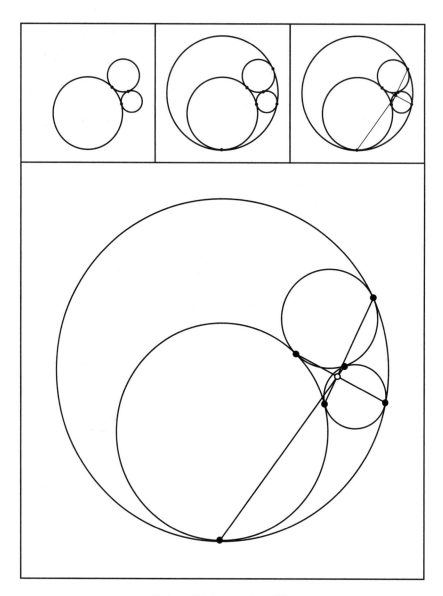

Point of intersection 37
Three circles in a circle

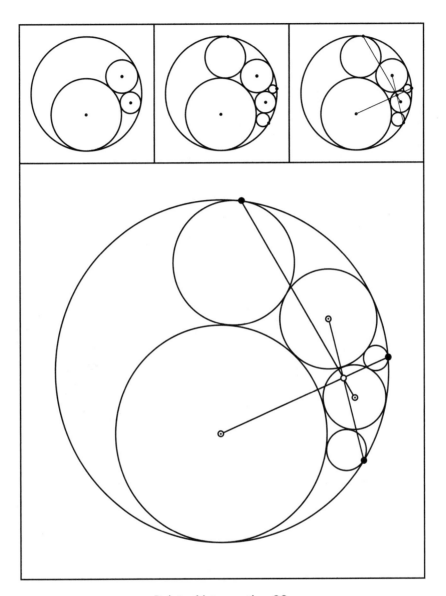

Point of intersection 38

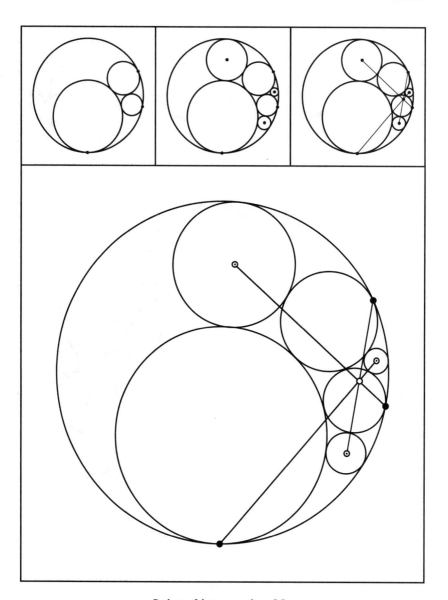

Point of intersection 39

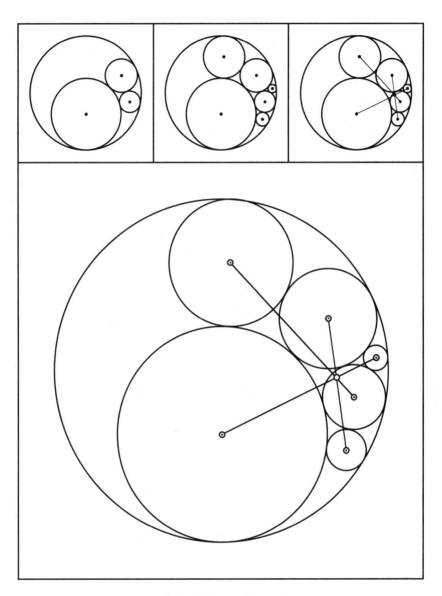

Point of intersection 40

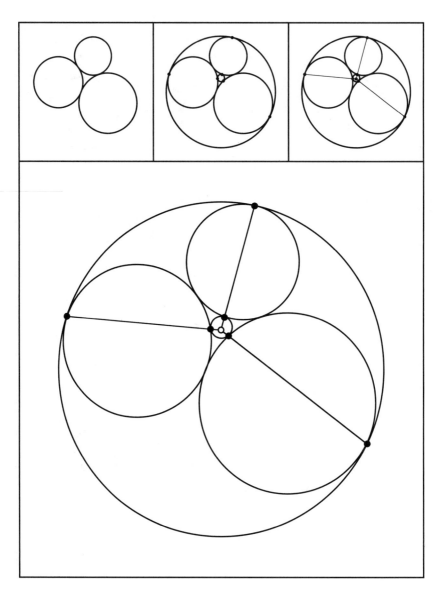

Point of intersection 41
See [Berger 1987], p. 317

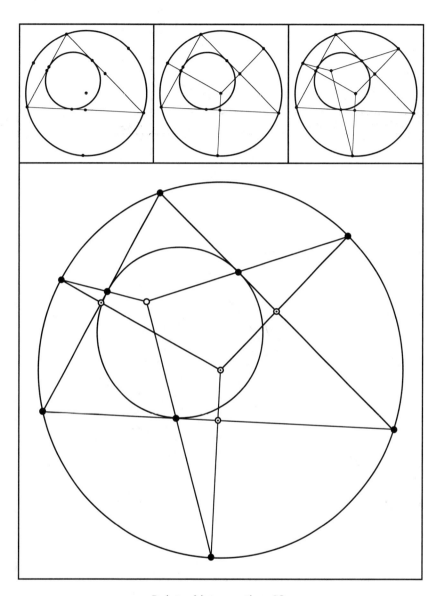

Point of intersection 42

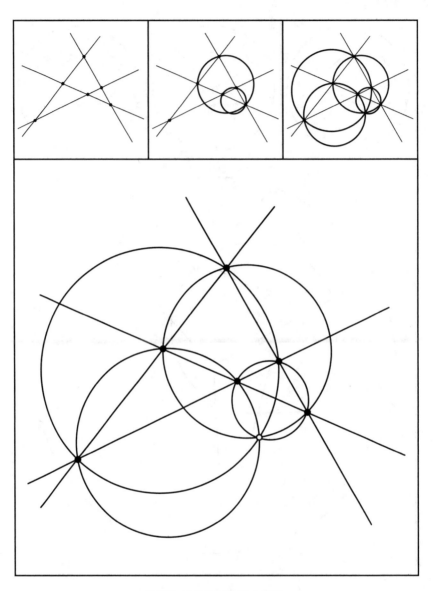

Point of intersection 43

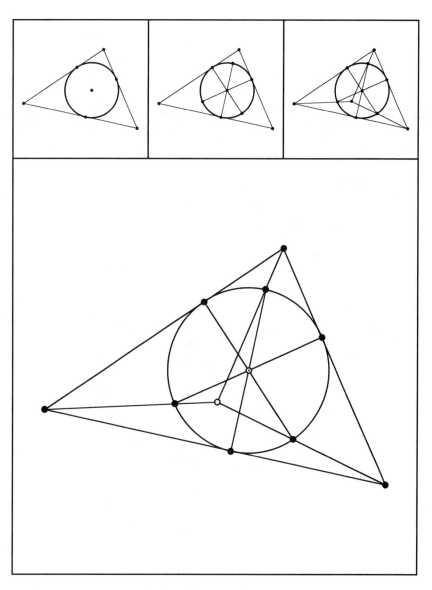

Point of intersection 44

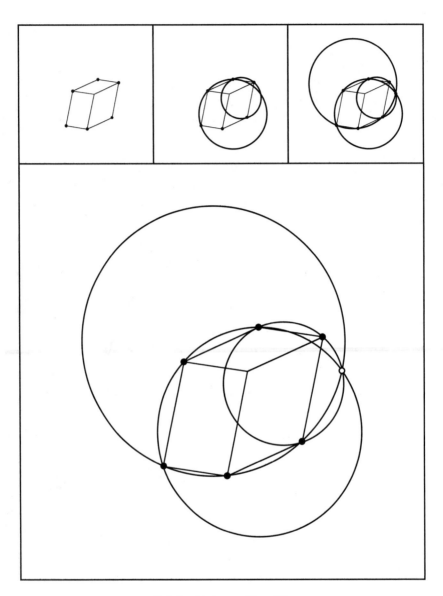

Point of intersection 45
"Cube" and circles

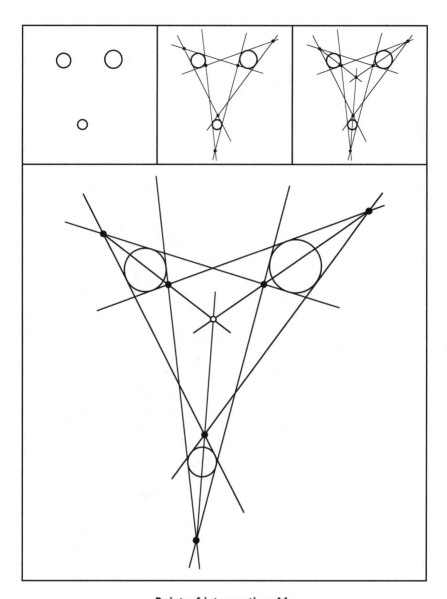

Point of intersection 46

Common tangents, see [Walser 1994]

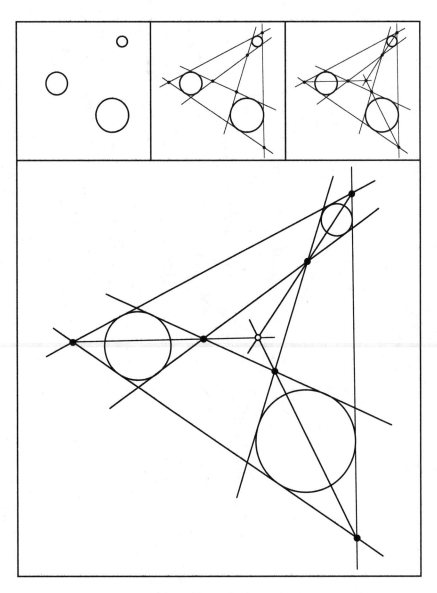

Point of intersection 47

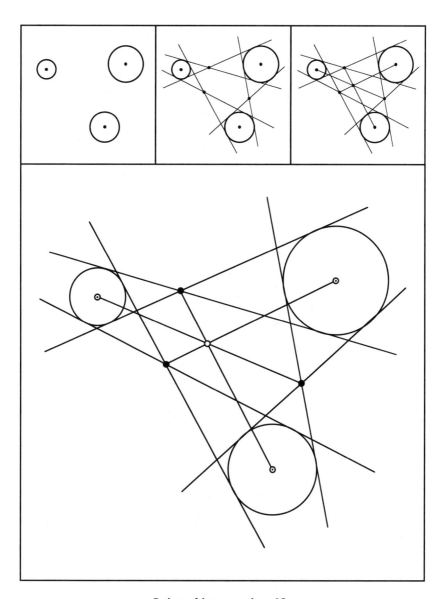

Point of intersection 48

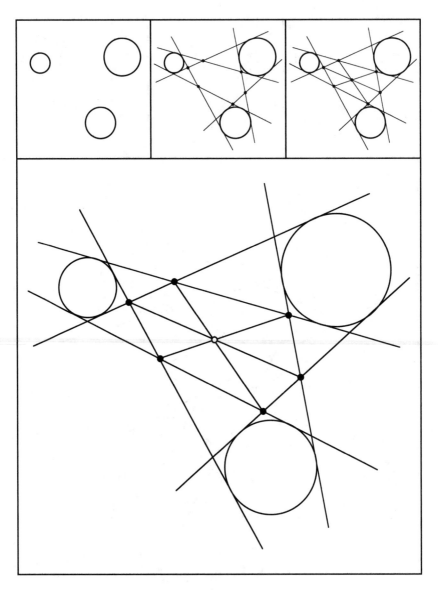

Point of intersection 49

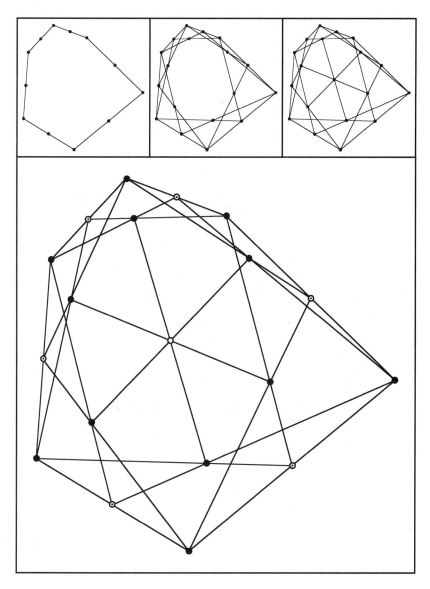

Point of intersection 50

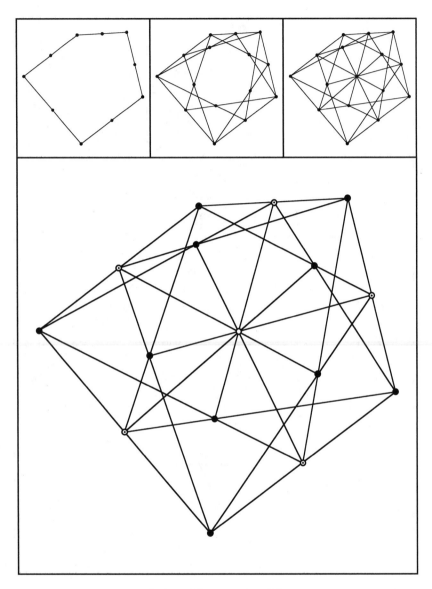

Point of intersection 51

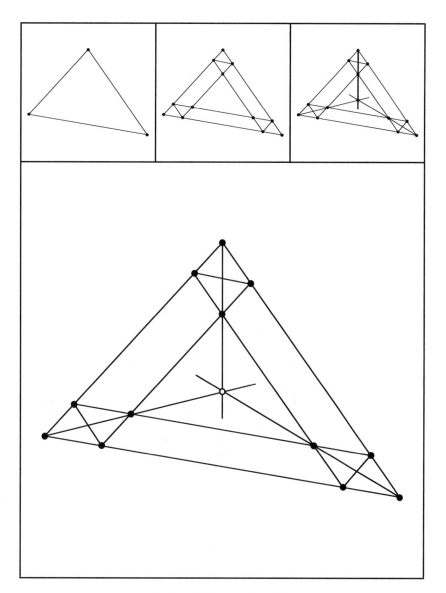

Point of intersection 52
Circuit, see [Kroll 1990]

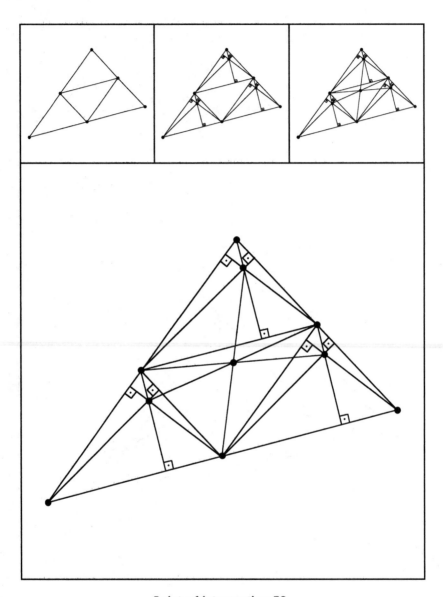

Point of intersection 53

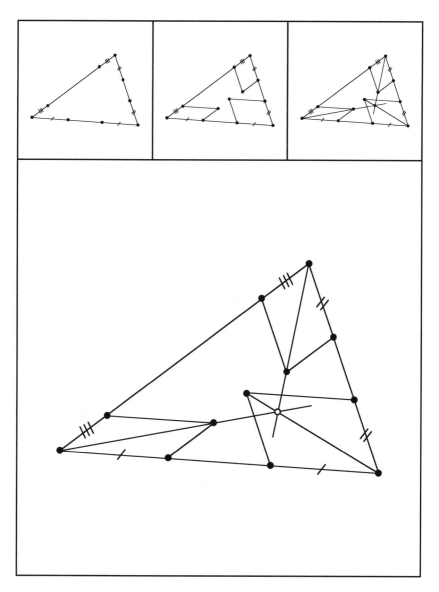

Point of intersection 54

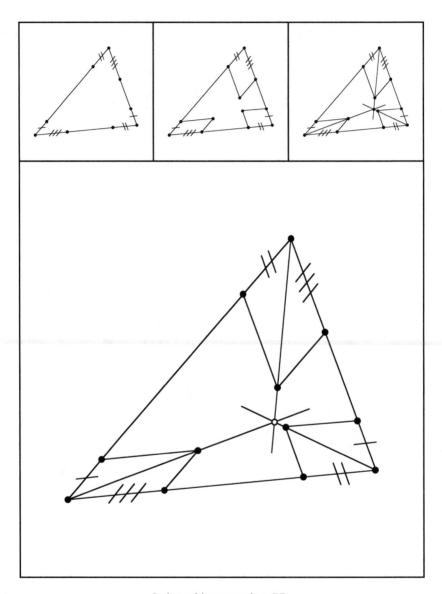

Point of intersection 55

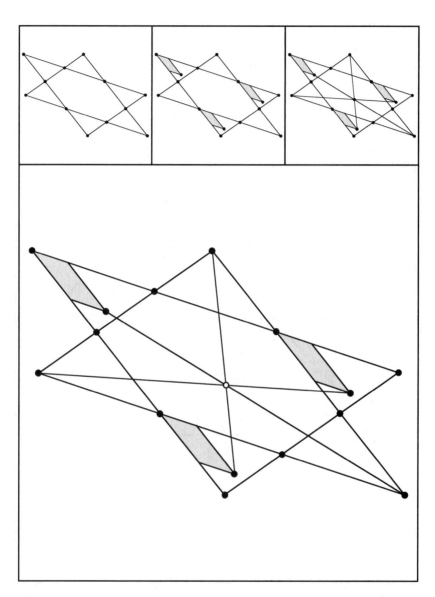

Point of intersection 56

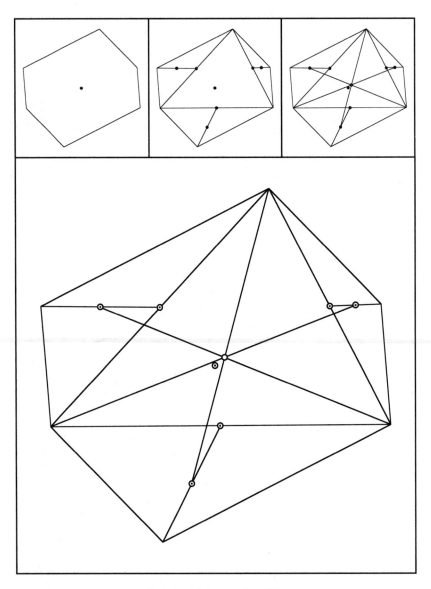

Point of intersection 57

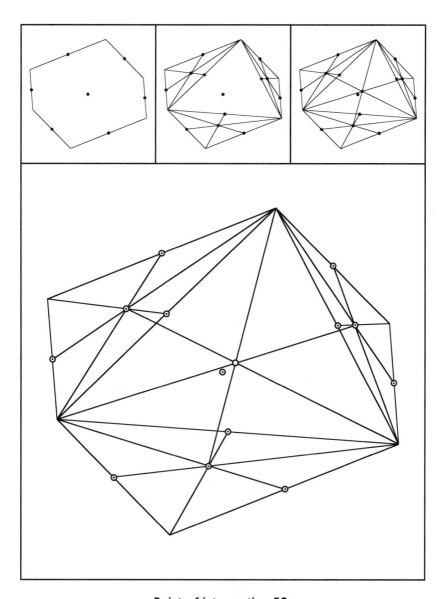

Point of intersection 58

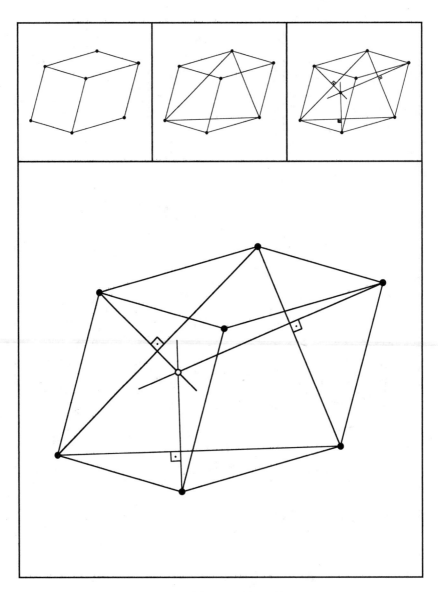

Point of intersection 59
"Cube"

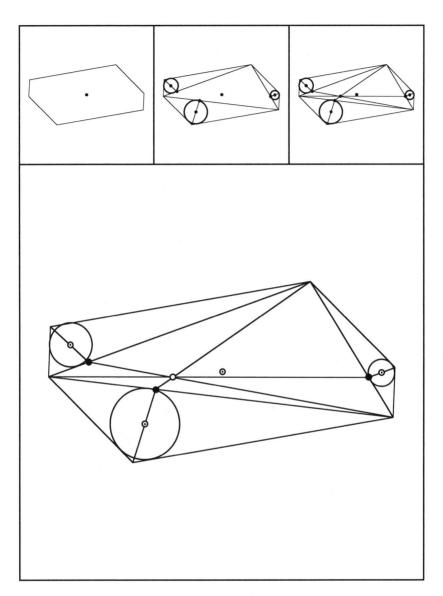

Point of intersection 60

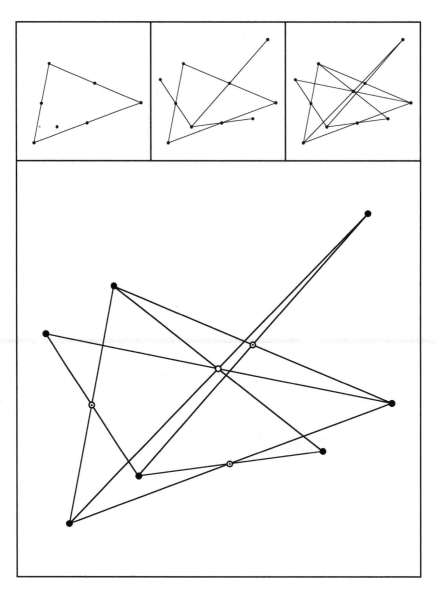

Point of intersection 61

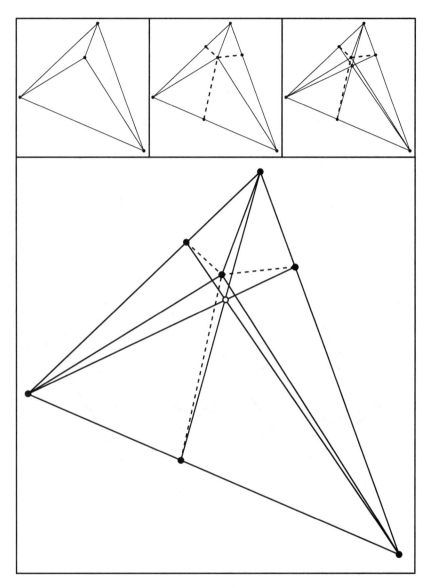

Point of intersection 62

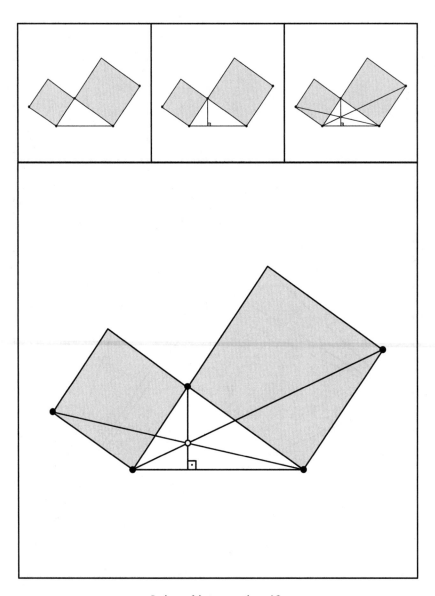

Point of intersection 63
Homage to Pythagoras

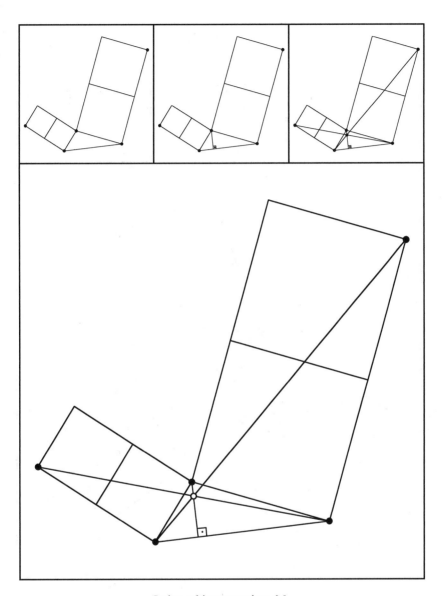

Point of intersection 64

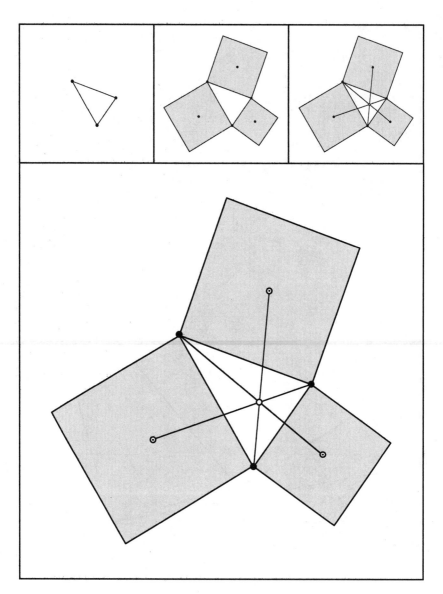

Point of intersection 65
No greetings from Pythagoras

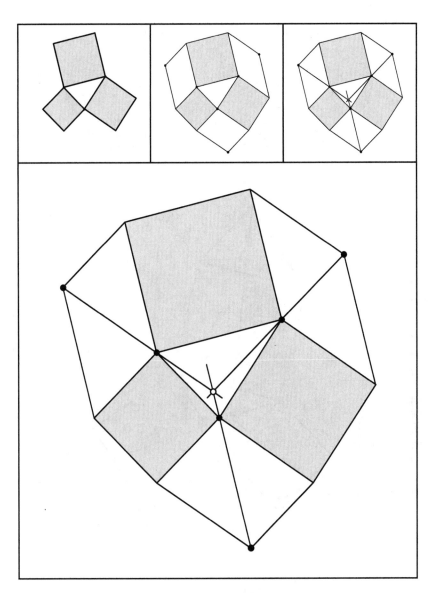

Point of intersection 66

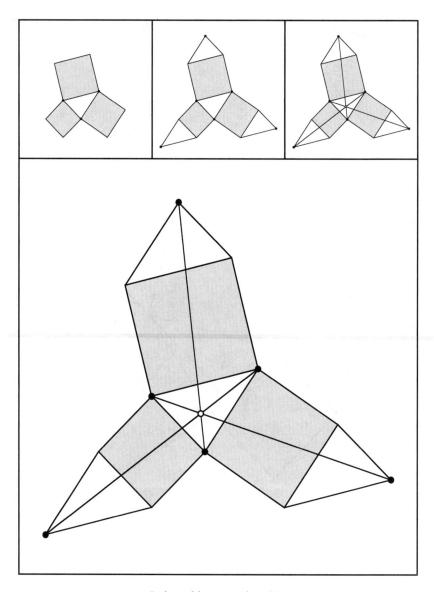

Point of intersection 67

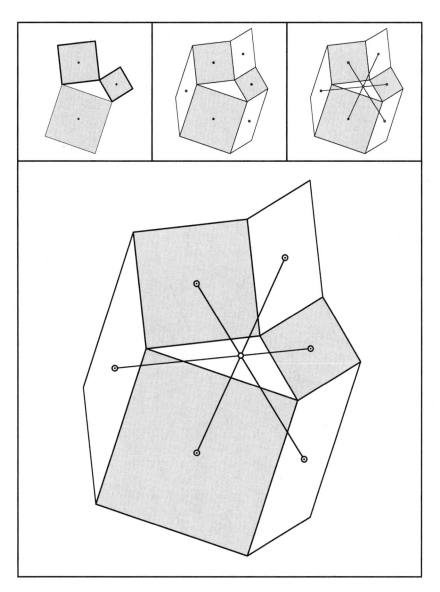

Point of intersection 68

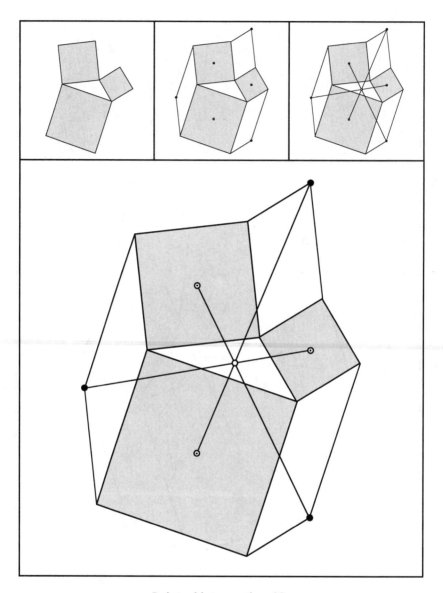

Point of intersection 69

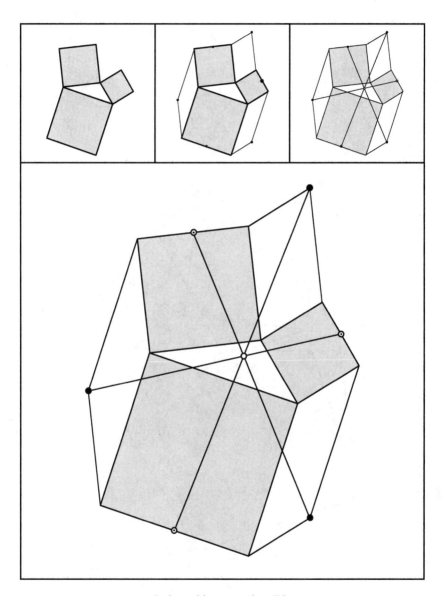

Point of intersection 70

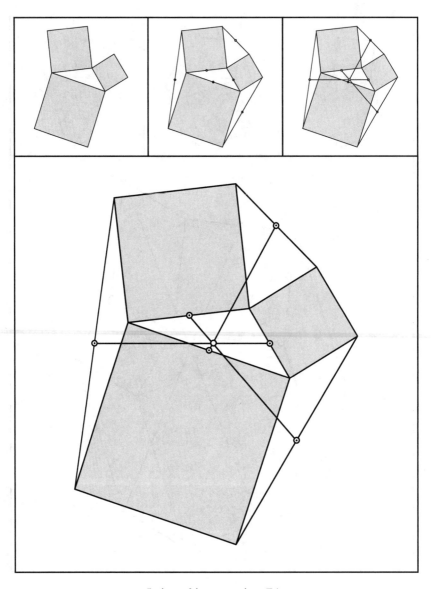

Point of intersection 71

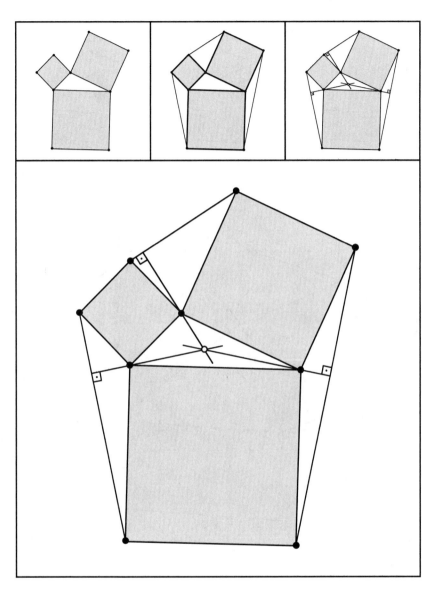

Point of intersection 72
See [Hoehn 2001]

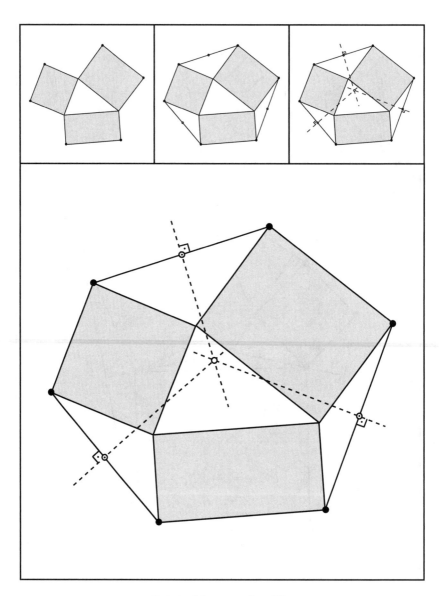

Point of intersection 73
See PM, Praxis der Mathematik, 3/39, p. 138, Problem 685

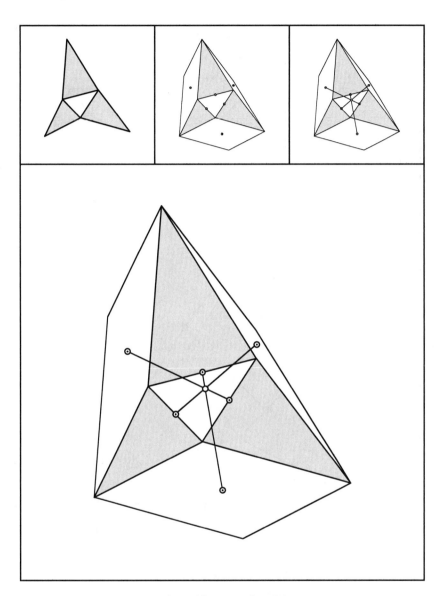

Point of intersection 74

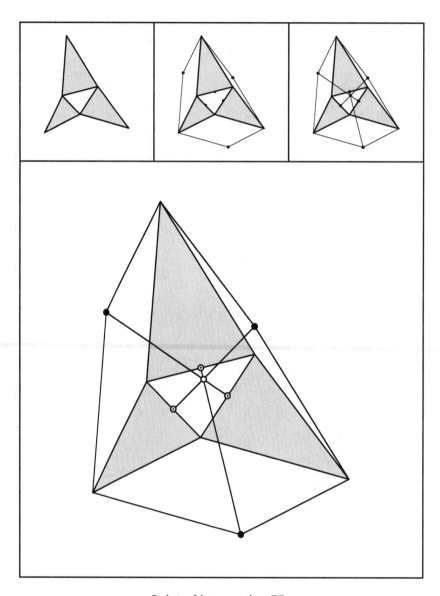

Point of intersection 75

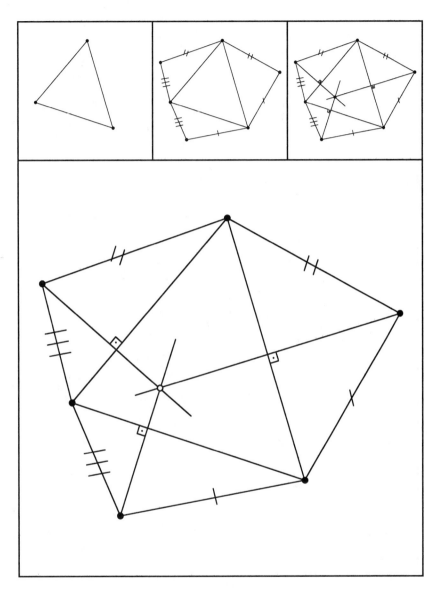

Point of intersection 76
Idea: Roland Wyss

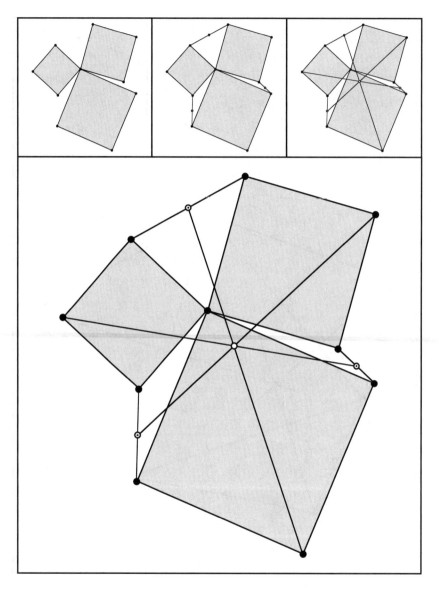

Point of intersection 77

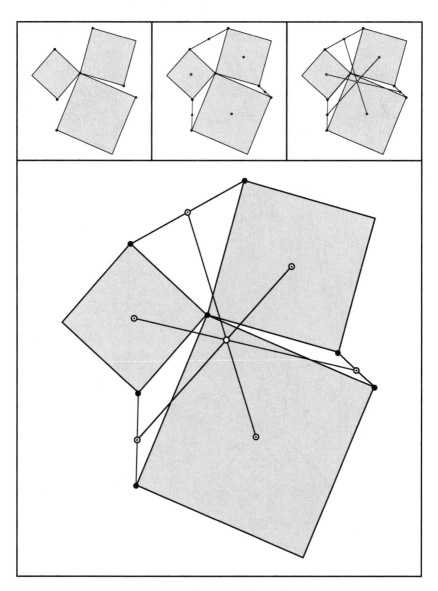

Point of intersection 78

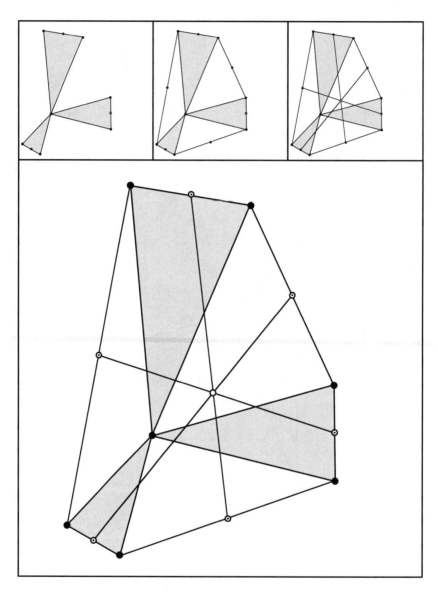

Point of intersection 79
Propeller

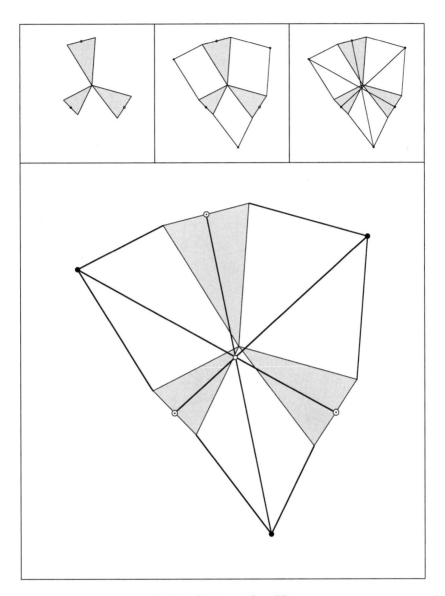

Point of intersection 80

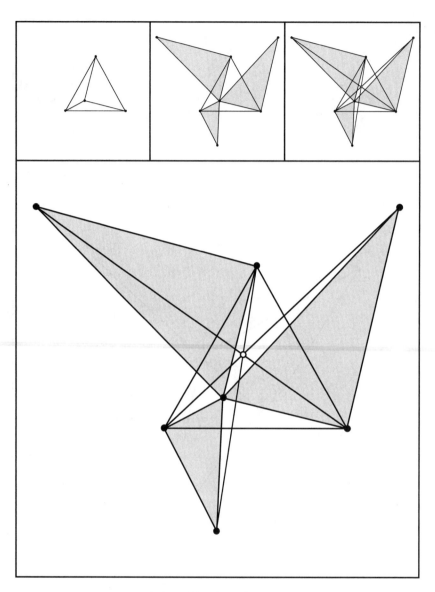

Point of intersection 81

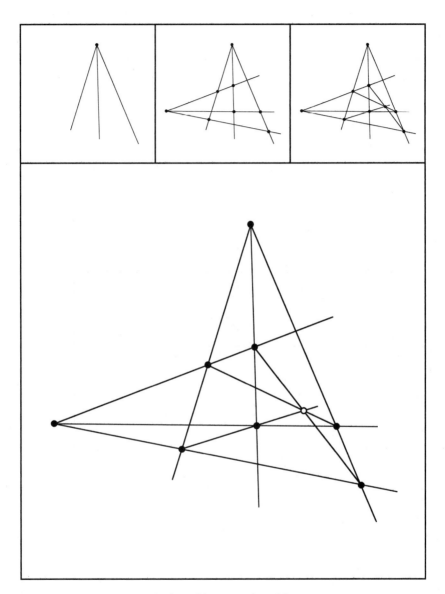

Point of intersection 82
Homage to Girard Desargues

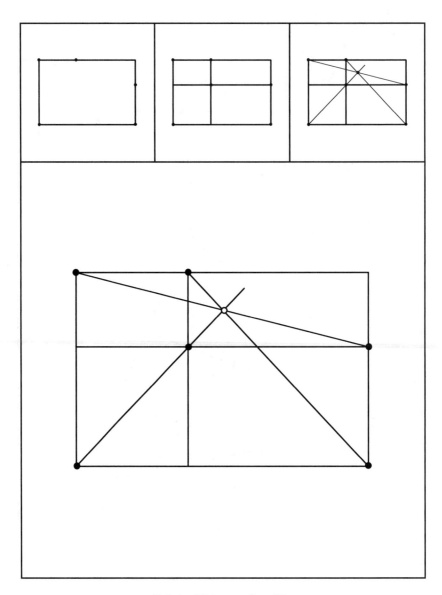

Point of intersection 83
See [Wells 1991], p. 169

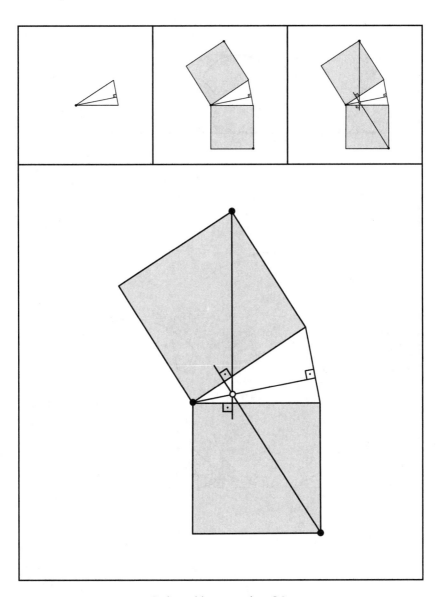

Point of intersection 84
See remarks 3.6.4

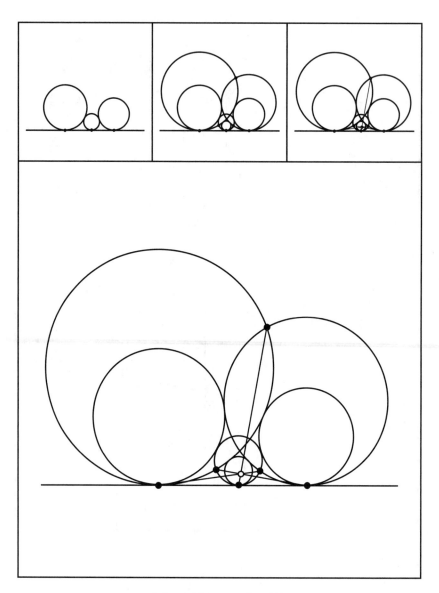

Point of intersection 85
Kissing circles

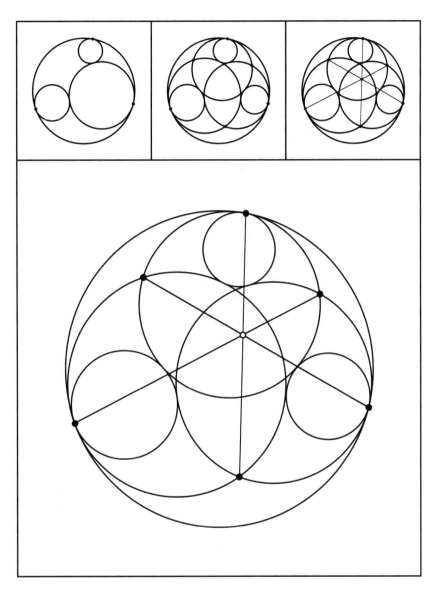

Point of intersection 86

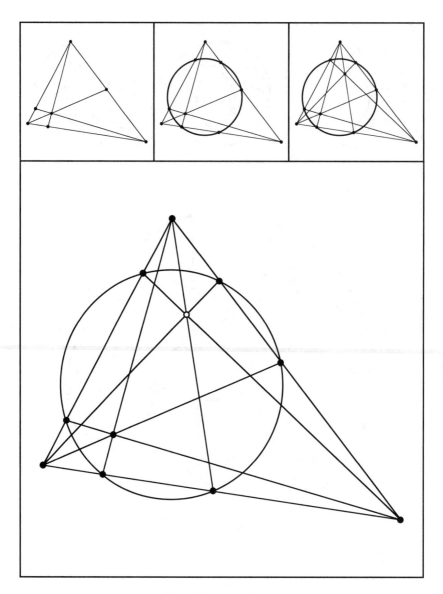

Point of intersection 87
Homage to Karl Feuerbach

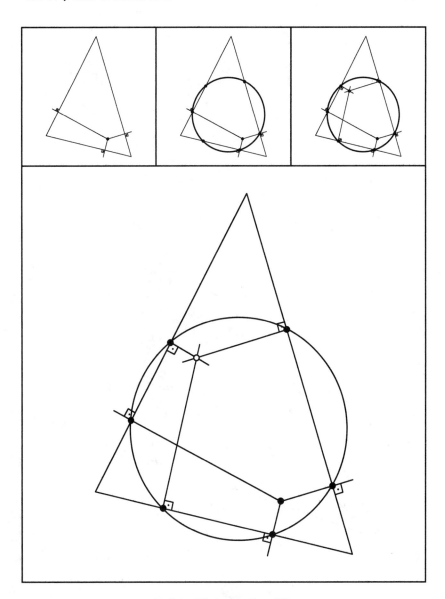

Point of intersection 88

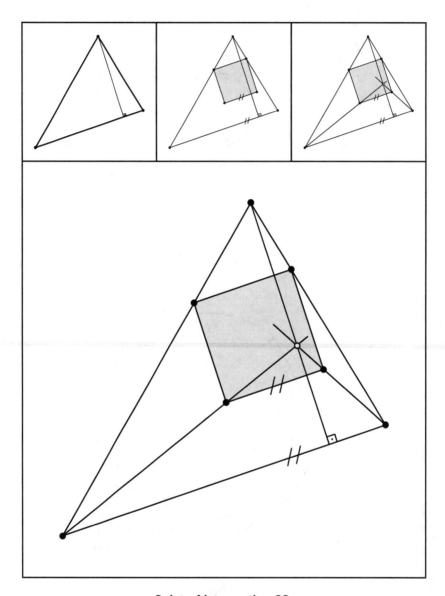

Point of intersection 89

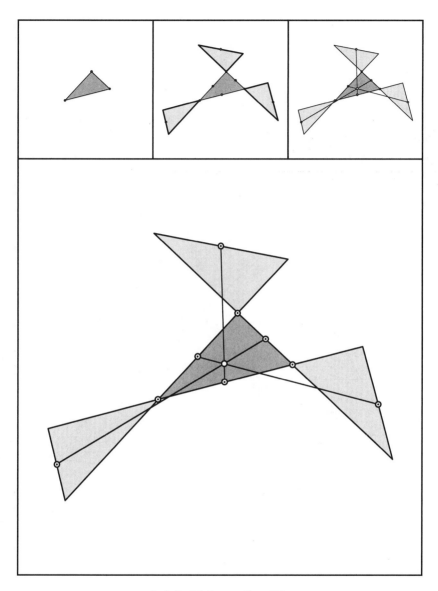

Point of intersection 90
Butterflies

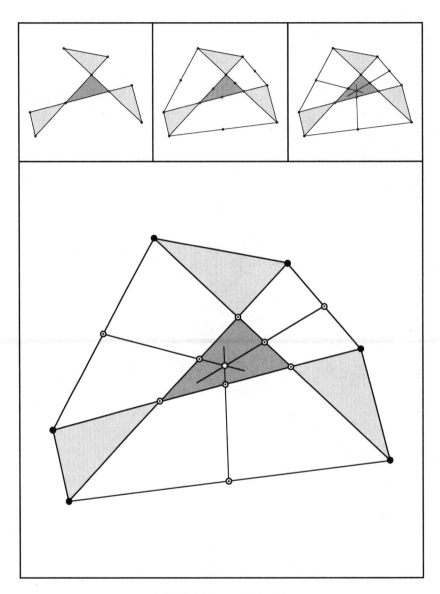

Point of intersection 91

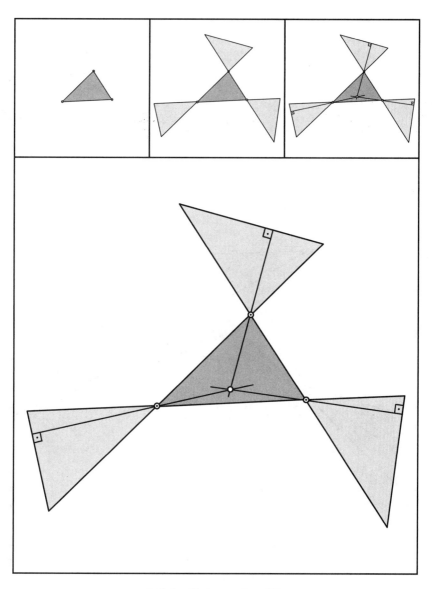

Point of intersection 92

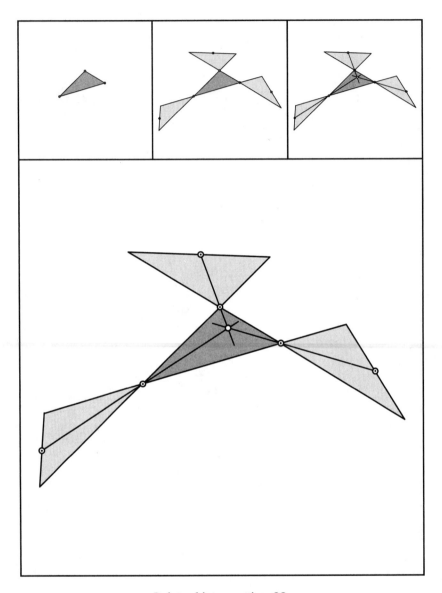

Point of intersection 93
J. T. Groenman, Problem 798, Elemente der Mathematik, 1978, p. 19

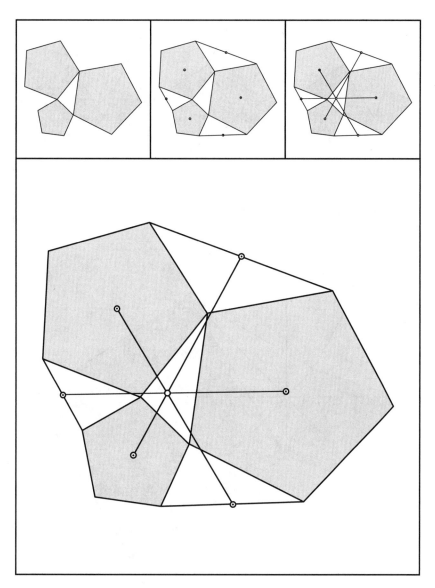

Point of intersection 94
Regular pentagons

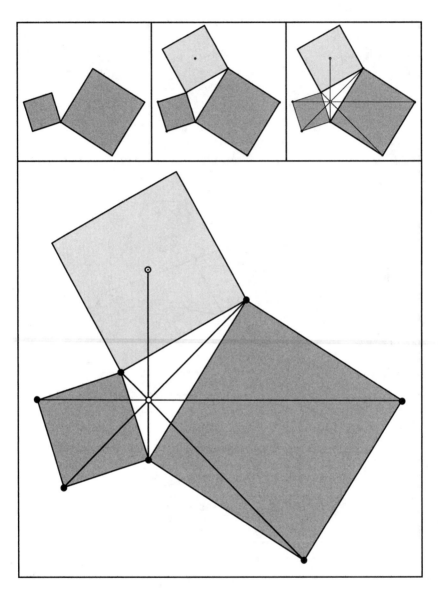

Point of intersection 95
Two squares with a third

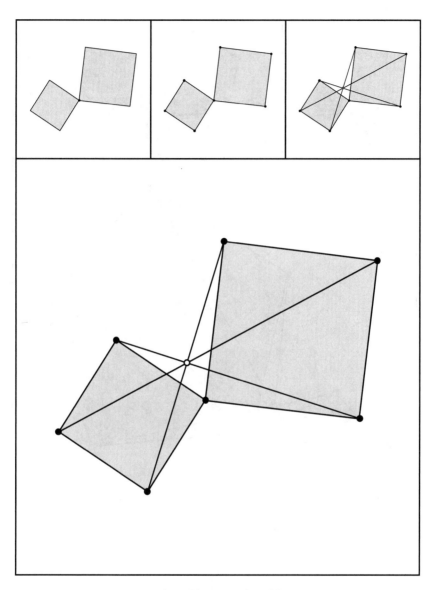

Point of intersection 96
See [Detemple/Harold 1996], p. 19, and Remarks 2.6.6

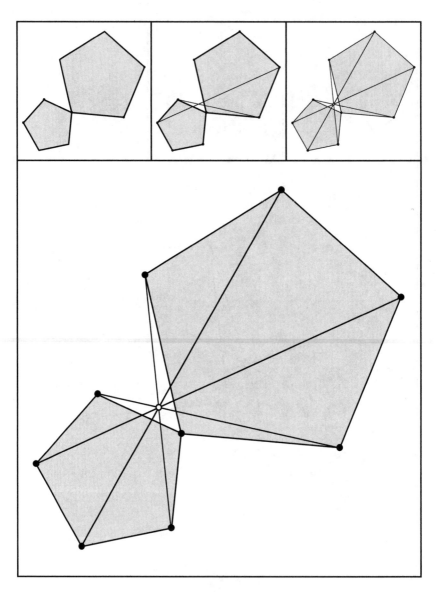

Point of intersection 97
Regular pentagons

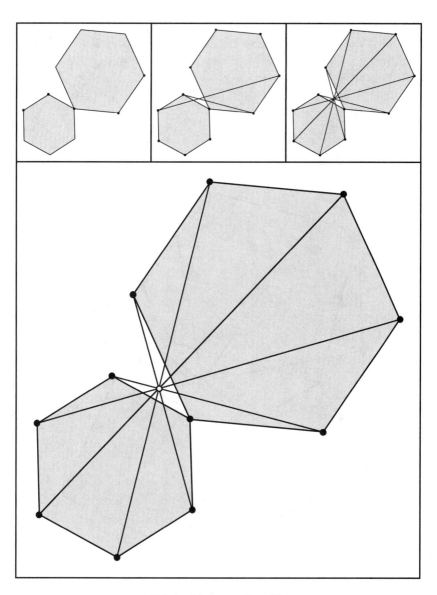

Point of intersection 98
Regular hexagons

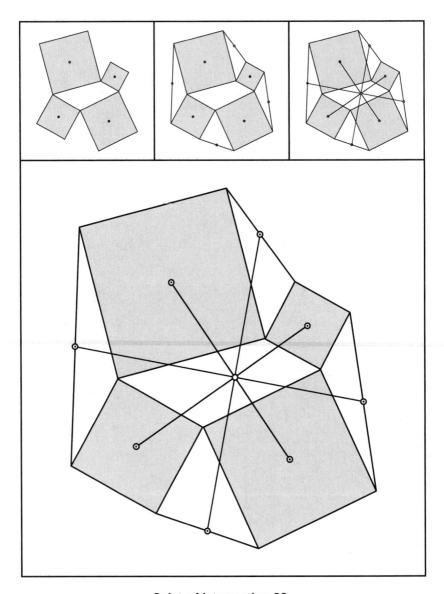

Point of intersection 99
See [Detemple/Harold 1996], p. 25

CHAPTER 3

The background

3.1 The four classical points of intersection

The four "classical" points of intersection are associated with points of intersection in the triangle: the *center of gravity* (or *centroid*) S (point of intersection of the medians), the point of intersection U of the perpendicular bisectors of the sides (*center of the circumcircle*), the point of intersection H of the altitudes, and the point of intersection I of the angle bisectors (*center of the incircle*) (see Figure 24). It turns out that S, H, and U lie on a line — the so-called *Euler line*. The incenter is exceptional.

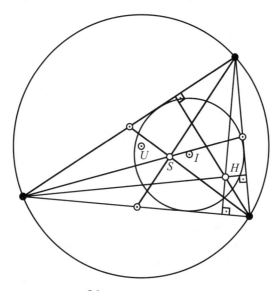

FIGURE 24
The four classical points of intersection

Besides these four classical examples, there are also many further points of intersection and special points in the triangle, see [Baptist 1992], [Donath 1976], [Hauptmann 1995], [Kimberling 1998], [Klemmenz 2003], [Longuet-Higgens 2001], [mathworld], [Walser 1990–1994], [Walser 1993].

3.2   Proof strategies

> *When shall we three meet again,*
> *In thunder, lightning or in rain?*
> Shakespeare, Macbeth

If three straight lines pass through a common point — three such straight lines are called concurrent — this may either be an accident or a special property of these three straight lines (for example, the three medians of a triangle), or a property that holds in every triangle. The question then naturally follows, how and why the property of concurrence holds for all triangles; in other words, how can this property be proved for an arbitrary triangle? Entirely different methods can be employed, as will be shown with the example of the centroid.

3.2.1   The classical proof: the dialogue

We place the thought processes of Petra and Quasi against each other. Petra argues thus (Figure 25):

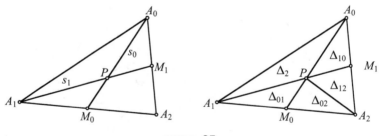

FIGURE **25**
Subdivision

In the triangle $A_0 A_1 A_2$ the two medians s_0 and s_1 intersect at P. With the line segment $\overline{A_2 P}$ there arise four small subtriangles Δ_{01}, Δ_{02}, Δ_{12}, Δ_{10}, as well as an obviously larger residual triangle Δ_2.

The triangles Δ_{01} and Δ_{02} have the same area, since they possess equal bases $\overline{A_1 M_0}$ and $\overline{M_0 A_2}$ and the same altitude from P. Analogously (that is, with the same thought process but with a different "identification" of the various geometrical entities involved) it follows that the triangles Δ_{12} and Δ_{10} have the same area.

But now, on the other hand, $\Delta_{01} + \Delta_{02} + \Delta_{12}$ occupies half of the area of the triangle (why?) and so does $\Delta_{02} + \Delta_{12} + \Delta_{10}$. It thus follows that $\Delta_{01} = \Delta_{10}$, so that all four small subtriangles have the same area. The subtriangle Δ_{12} therefore has one third of the area of the triangle $A_1 A_2 M_1$. Since these two triangles share the altitude from A_2 it follows that $\overline{M_1 P}$ is one third of the length of the segment $\overline{M_1 A_1}$. Petra also shows, analogously, that the segment $\overline{M_0 P}$ measures one third of the segment $\overline{M_0 A_0}$. So much for Petra.

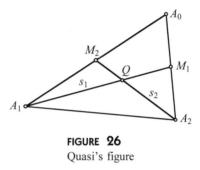

FIGURE 26
Quasi's figure

Quasi proceeds from a different basic figure: He has the two medians s_1 and s_2 intersecting in Q.

His thinking, analogous to that of Petra, leads to the following conclusions: The segment $\overline{M_1 Q}$ measures one third of the segment $\overline{M_1 A_1}$, and the segment $\overline{M_2 Q}$ measures one third of $\overline{M_2 A_2}$.

Now we seem to have two mutually contradictory statements: Petra finds that the segment $\overline{M_1 P}$ measures one third of the segment $\overline{M_1 A_1}$. Quasi on the other hand finds that the segment $\overline{M_1 Q}$ measures one third of the segment $\overline{M_1 A_1}$. Since both P and Q lie in the interior of the segment $\overline{M_1 A_1}$, we can only avoid a contradiction if P and Q are the same point. Thus it is also clear that all three medians pass through this point.

3.2.2 Proofs by calculation

Straight lines and their points of intersection may be represented with the help of vector geometry. We denote by $\overline{a_i}$ the vector of coordinates of the point A_i, and by $\overline{m_i}$ the vector of coordinates of the midpoint M_i of a side. We then have (the indices are always to be understood modulo 3):

$$\overline{m_i} = \frac{1}{2} (\overline{a}_{i+1} + \overline{a}_{i+2}) .$$

For the median s_i we have the parametric representation

$$s_i: \quad \overline{x}_i(t_i) = \frac{1}{2}(\overline{a}_{i+1} + \overline{a}_{i+2}) + t_i\left(\overline{a}_i - \frac{1}{2}(\overline{a}_{i+1} + \overline{a}_{i+2})\right).$$

For $t_i = \frac{1}{3}$ this yields $\overline{x}_i\left(\frac{1}{3}\right) = \frac{1}{3}(\overline{a}_0 + \overline{a}_1 + \overline{a}_2)$. This result is independent of the index i, that is to say, all three medians pass through this point.

In our calculations we have chosen the vertices of the triangle $A_0 A_1 A_2$ arbitrarily. Sometimes, however, it is meaningful to make a very special choice. Since intersection problems are invariant under similarity transformations, the length of one side of the triangle can be given in advance. A particular special case for coordinates of the vertices is, for example, $A_0(0,0)$, $A_1(1,0)$, $A_2(p,q)$ (Figure 27). The entire configuration then depends on the two parameters p and q.

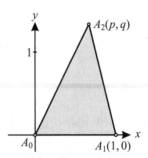

FIGURE 27
Special choice of coordinates

Thus we can carry out the following "rapid-fire calculations": We obtain the following explicit equations for the medians

$$s_0: \quad y = \frac{q}{1+p}x$$

$$s_1: \quad y = \frac{q}{p-2}x - \frac{q}{p-2}$$

$$s_2: \quad y = \frac{2q}{2p-1}x - \frac{q}{2p-1}.$$

For the point of intersection of s_0 and s_1 we set $\frac{q}{1+p}x = \frac{q}{p-2}x - \frac{q}{p-2}$. After a little algebra we obtain: $x = \frac{1+p}{3}$. The height q obviously plays no role in determining the x-coordinate. To find the y-coordinate, we substitute $x = \frac{1+p}{3}$ in the equation for s_0 and obtain $y = \frac{q}{3}$. The point of intersection of s_0 and s_1 thus has coordinates $\left(\frac{1+p}{3}, \frac{q}{3}\right)$. We now verify that this point also lies on the median s_2.

The algebraic calculations in such a proof can be complicated and are in practice often only carried out with the help[1] of a Computer-Algebra System (CAS) like, for example, *Maple* or *Mathematica*. For orthodox geometers the question then immediately presents itself, whether such a computer proof really is a "valid" proof. Some of the 99 points of intersection have only been "proved" using such a CAS.

3.2.3   Dynamic Geometry Software

In the United States commercial Dynamic Geometry Software (DGS) such as Cabri Geometry and Geometer's Sketchpad possess as a rule a "moving mode".[2] In a completed construction the initial data, for example the three vertices of a triangle basic to the construction, can be altered subsequently by moving a point using the mouse (see [Schumann 1990/91]). A generally valid point of intersection of three straight lines remains a point of intersection under this alteration process. The question now is, conversely, whether the invariance of a (hypothetical) point of intersection in the moving mode can be regarded as a "proof." Only finitely many cases can be considered — in view of the pixel density. On the other hand, the probability that the property of being a point of intersection is accidental is so small that we must become accustomed to adopting a tolerant approach towards moving mode "proofs," even computer proofs. We recognize the danger of having "almost points of intersection," where three straight lines do not intersect in a point but, throughout the moving process, form a triangle that can be covered by a circular disc of constant ("small") radius. *Cabri Geometry I* indeed offers the possibility of magnification, but that has an upper limit. In any case, the moving mode is a good interactive instrument for the discovery, under given initial conditions, of a point of intersection.

Dynamic Geometry Software usually offers also the possibility of raising the question of the incidence of a point with a straight line. The answer depends on an underlying Computer-Algebra System. Here too, naturally, there occurs the question of the validity of a computer proof.

---

[1] Translators' note: Such a CAS is not needed for the calculation immediately above.

[2] Programs, in English, such as Cinderella and Euklid, may be found, free of charge, at [mathforum].

3.2.4 Affine invariance

Geometric constructions, which are based only on incidence, parallelism, and the ratio of parts (for example, midpoints), are invariant under affine maps. An example of this is the point of intersection of the medians, that is, the centroid. Since any arbitrary triangle can be regarded as the affine image of an equilateral triangle, it suffices in such cases to prove the concurrence property for equilateral triangles, where the proof is really simple (see Figure 28).

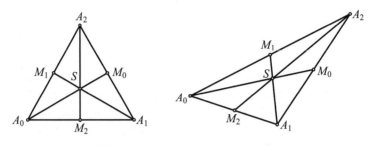

FIGURE 28
A triangle as the affine image of an equilateral triangle

This procedure does not work if right angles or bisectors of angles are in question, as in the intersection of the altitudes, the center of the circumcircle, and the center of the incircle.

3.3 Central projection

Two figures which can be mapped on each other by a central projection are called *perspectively similar* (*homothetic*) (Figure 29); moreover, such corresponding straight lines are parallel.

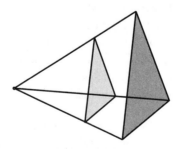

FIGURE 29
Perspective similarity

Conversely, the straight lines joining corresponding points of two perspectively similar triangles meet in a point; any two intersect in the center of projection. Once again the simplest example is the centroid: the two triangles $A_0 A_1 A_2$ and $M_0 M_1 M_2$ are perspectively similar (Figure 30).

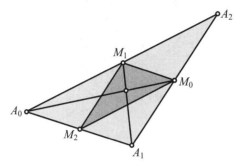

FIGURE 30
Perspective similarity with respect to the centroid

As a further example, we start with a "circuit" in a triangle, which begins at an arbitrary point on a side of the triangle and consists of line segments parallel to the sides of the triangle. The path comes to an end after six steps (Figure 31), making a circuit (see [Kroll 1990]), and divides the original triangle into several pieces.

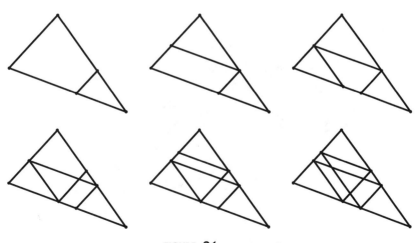

FIGURE 31
Circuit in six steps

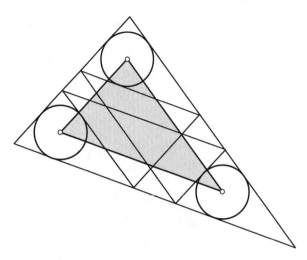

FIGURE 32
Perspectively similar triangles

The three centers of the incircles drawn in the outermost subtriangles (Figure 32) form a triangle perspectively similar to the original triangle.

But the same also holds for the centers of the additional three incircles drawn in Figure 33. The straight lines joining corresponding points are thus concurrent (Point of intersection 18). It follows readily that many other points

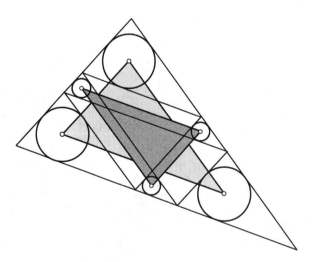

FIGURE 33
A further perspectively similar triangle

of intersection may be found and proved in this way (see Points of intersection 16, 17, 19, 20, 21, 27, 52, 53, 56).

3.4 Ceva's Theorem

3.4.1 Giovanni Ceva

Ceva's Theorem is a very efficient tool for proving points-of-intersection results. It is also very interesting from a historic standpoint. It is the first theorem of elementary geometry that was not already known to the Greeks. Giovanni Ceva (1647–1734) lived in Mantua and, in 1678 in Milan, published the monograph *De lineis rectis se invicem secantibus, statica constructio*. Ceva considered a triangle with unequal weights placed at its vertices and asked where their center of gravity would be (see [Chasles 1968]). The usual centroid of a triangle appears in these considerations as the special case of equal weight distribution 1 : 1 : 1. Ceva's Theorem asserts the following (Figure 34):

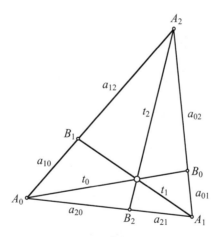

FIGURE 34
Ceva's Theorem

Three vertex-transversals t_0, t_1, t_2 are concurrent if and only if the ratios induced on the opposite sides satisfy the condition:

$$\frac{a_{01}}{a_{02}} \ \frac{a_{12}}{a_{10}} \ \frac{a_{20}}{a_{21}} = 1.$$

Remark. The ratios $a_{i,i+1}/a_{i,i+2}$ appearing in Ceva's Theorem are very often furnished with a sign, which is negative precisely when the point B_i is interior to the segment $\overline{A_{i+1}A_{i+2}}$. In this notation the product of the three ratios must be -1. If the product is $+1$, then the three points B_0, B_1, and B_2 are collinear (Theorem of Menelaus). Ceva's Theorem remains valid if the common point of the three vertex-transversals lies outside the triangle. The concepts and ratios appearing in the statement of Ceva's Theorem are affine invariants.

For a proof of Ceva's Theorem we leave the three concurrent vertex-transversals and exploit the areas of certain subtriangles (Figure 35).

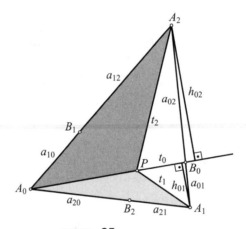

FIGURE 35
Area ratios of subtriangles

The two triangles $A_0 A_1 P$ and $A_0 A_2 P$ have a common side $\overline{PA_0}$. Their areas are thus in the same ratio as the appropriate altitudes h_{01} and h_{02}. From the relevant theorems on similar triangles, these two altitudes h_{01} and h_{02} are in the same ratio as the two segments a_{01} and a_{02}. Thus, with A standing for area, we have:

$$\frac{a_{01}}{a_{02}} = \frac{A_{\triangle A_0 A_1 P}}{A_{\triangle A_0 A_2 P}}.$$

Likewise we have

$$\frac{a_{12}}{a_{10}} = \frac{A_{\triangle A_1 A_2 P}}{A_{\triangle A_1 A_0 P}} \quad \text{and} \quad \frac{a_{20}}{a_{21}} = \frac{A_{\triangle A_2 A_0 P}}{A_{\triangle A_2 A_1 P}}.$$

From this it follows that

$$\frac{a_{01}}{a_{02}} \frac{a_{12}}{a_{10}} \frac{a_{20}}{a_{21}} = 1.$$

For three vertex-transversals which are not concurrent (Figure 36), we consider the point of intersection of two of the three vertex-transversals, for example, P_{01} as the point of intersection of t_0 and t_1.

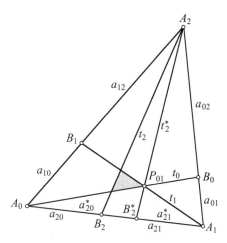

FIGURE 36
Non-concurrent vertex-transversals

Let t_2^* be the transversal through A_2 and P_0. Thus we have

$$\frac{a_{01}}{a_{02}} \frac{a_{12}}{a_{10}} \frac{a_{20}^*}{a_{21}^*} = 1.$$

Since $a_{20} < a_{20}^*$ and $a_{21} > a_{21}^*$ (or, conversely, $a_{20} > a_{20}^*$ and $a_{21} < a_{21}^*$), it follows that

$$\frac{a_{20}^*}{a_{21}^*} \neq \frac{a_{20}}{a_{21}}, \quad \text{and so} \quad \frac{a_{01}}{a_{02}} \frac{a_{12}}{a_{10}} \frac{a_{20}}{a_{21}} \neq 1.$$

This completes the proof of Ceva's Theorem.

3.4.2   Examples

3.4.2.1   The center of gravity. The points B_i are the midpoints of the segments $\overline{A_{i+1} A_{i+2}}$. Thus

$$\frac{a_{01}}{a_{02}} = 1, \quad \frac{a_{12}}{a_{10}} = 1, \quad \frac{a_{20}}{a_{21}} = 1,$$

whence, immediately,

$$\frac{a_{01}}{a_{02}} \frac{a_{12}}{a_{10}} \frac{a_{20}}{a_{21}} = 1.$$

3.4.2.2   The point of intersection of the altitudes.   From $A := A_{\triangle A_0 A_1 A_2} = \frac{1}{2} h_0 a_0$ it follows that $h_0 = \frac{2A}{a_0}$.

By Pythagoras' Theorem we then have:

$$\frac{a_{01}}{a_{02}} = \frac{\sqrt{a_2^2 - h_0^2}}{\sqrt{a_1^2 - h_0^2}} = \frac{\sqrt{a_2^2 - \frac{4A^2}{a_0^2}}}{\sqrt{a_1^2 - \frac{4A^2}{a_0^2}}} = \frac{\sqrt{a_0^2 a_2^2 - 4A^2}}{\sqrt{a_0^2 a_1^2 - 4A^2}}.$$

Similarly

$$\frac{a_{12}}{a_{10}} = \frac{\sqrt{a_1^2 a_0^2 - 4A^2}}{\sqrt{a_1^2 a_2^2 - 4A^2}} \quad \text{and} \quad \frac{a_{20}}{a_{21}} = \frac{\sqrt{a_2^2 a_1^2 - 4A^2}}{\sqrt{a_2^2 a_0^2 - 4A^2}}.$$

From this it follows that

$$\frac{a_{01}}{a_{02}} \frac{a_{12}}{a_{10}} \frac{a_{20}}{a_{21}} = 1,$$

so that the existence of the point of intersection of the altitudes is ensured.

Remark.   The existence of the point of intersection of the altitudes may be proved without using Ceva's Theorem. The point of intersection of the

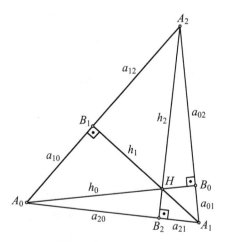

FIGURE 37
The point of intersection of the altitudes

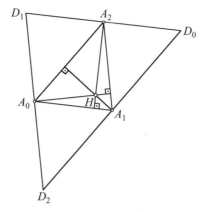

FIGURE **38**
Point of intersection of the altitudes as point of intersection of the right bisectors

altitudes of the triangle $A_0 A_1 A_2$ is, in fact, the point of intersection of the right bisectors of the sides of the triangle $D_0 D_1 D_2$, similar to the triangle $A_0 A_1 A_2$ but with sides twice the length (Figure 38).

3.4.3 The angle version of Ceva's Theorem

Instead of working with the segments cut off on the opposite side of the triangle, one can use the subdivisions of the angles of the triangle (Figure 39).

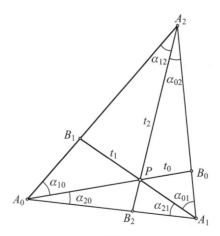

FIGURE **39**
Angle subdivisions

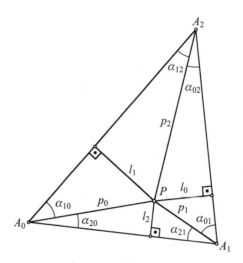

FIGURE 40
The designations

In the following arguments let p_i be the length of the segment $\overline{A_i P}$ and let l_i be the length of the perpendicular from P onto the side $a_i = \overline{A_{i+1} A_{i+2}}$ (Figure 40).

With these designations we have $l_0 = p_1 \sin \alpha_{01} = p_2 \sin \alpha_{02}$, whence

$$\frac{\sin \alpha_{01}}{\sin \alpha_{02}} = \frac{p_2}{p_1}.$$

Similarly

$$\frac{\sin \alpha_{12}}{\sin \alpha_{10}} = \frac{p_0}{p_2} \quad \text{and} \quad \frac{\sin \alpha_{20}}{\sin \alpha_{21}} = \frac{p_1}{p_0}.$$

This yields the angle version of Ceva's Theorem:

$$\frac{\sin \alpha_{01}}{\sin \alpha_{02}} \frac{\sin \alpha_{12}}{\sin \alpha_{10}} \frac{\sin \alpha_{20}}{\sin \alpha_{21}} = 1.$$

From this follows immediately, for example, the concurrence of the angle bisectors of a triangle, since, in this case, $\alpha_{10} = \alpha_{20}$, $\alpha_{21} = \alpha_{01}$, and $\alpha_{02} = \alpha_{12}$.

3.4.4 Generalization of the angle version

3.4.4.1 General n-gons. The angle version of Ceva's Theorem is also valid —in only one direction, however—for general n-gons $A_0, A_1, \ldots, A_{n-1}$, $n \geq 3$.

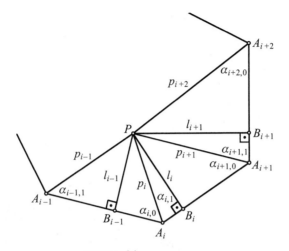

FIGURE 41
Designations in the n-gon

With the designations of Figure 41 we have

$$l_i = p_i \sin \alpha_{i,1} = p_{i+1} \sin \alpha_{i+1,0},$$

whence

$$\frac{\sin \alpha_{i,1}}{\sin \alpha_{i+1,0}} = \frac{p_{i+1}}{p_i}.$$

Thus

$$\prod_{i=0}^{n-1} \frac{\sin \alpha_{i,1}}{\sin \alpha_{i+1,0}} = \prod_{i=0}^{n-1} \frac{p_{i+1}}{p_i} = 1.$$

However, the converse is false, as the example of the angle bisectors in a rectangle already shows.

3.4.4.2 Spherical triangles. In spherical geometry great circles play the role of straight lines; a spherical triangle is thus bounded by three arcs of great circles. Three vertex transversals, which are now also arcs of great circles, are then concurrent if and only if

$$\frac{\sin \alpha_{01} \sin \alpha_{12} \sin \alpha_{20}}{\sin \alpha_{02} \sin \alpha_{10} \sin \alpha_{21}} = 1.$$

For a nice proof one works with the planes defined by the great circles and exploits their normal vectors. The angles referred to in Figure 42 are also then angles between these normal vectors, and the concurrence means that

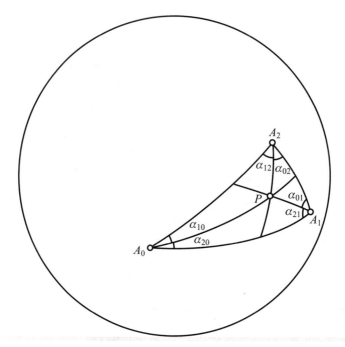

FIGURE 42
A spherical triangle on a sphere

the transversal planes have a common straight line intersection. But then their normal vectors are coplanar, so their determinant vanishes.

3.5   Jacobi's Theorem

3.5.1   A general theorem about points of intersection

We prove next a somewhat more general theorem, from which Jacobi's Theorem follows as a special case (see [Walser 1991]). On the sides $A_{i+1}A_{i+2}$, $i \in \{0, 1, 2\}$, of a triangle $A_0 A_1 A_2$, triangles $C_i A_{i+1} A_{i+2}$ and $D_i A_{i+1} A_{i+2}$ are constructed with the angles

$$\angle C_{i-1} A_i A_{i+1} = \angle C_{i+1} A_{i+1} A_{i+2} = \gamma_i$$

and

$$\angle D_{i-1} A_i A_{i+1} = \angle D_{i+1} A_i A_{i-1} = \delta_i \quad \text{(Figure 43)}.$$

For each $i \in \{0, 1, 2\}$ the angles γ_i and δ_i appear twice and have a common apex A_i. Further let B_i be the point of intersection of the straight line $C_i D_i$

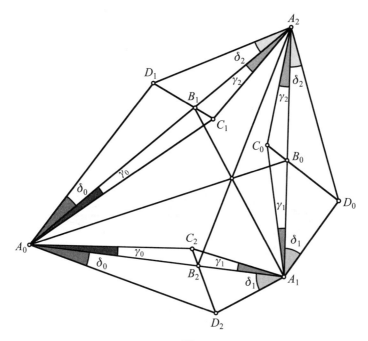

FIGURE 43
The starting figure

with the side $A_{i+1}A_{i+2}$ of the triangle. Then it follows that the three vertex transversals $A_i B_i$ are concurrent.

To prove this, we make next a "forward incision" in each of the triangles $A_1 A_2 C_0$ and $A_1 A_2 D_0$ with the common base $a_0 = \overline{A_1 A_2}$ (Figure 44).

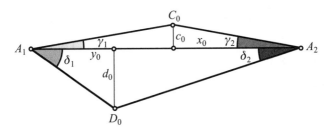

FIGURE 44
Forward incision

From $c_0 = x_0 \tan \gamma_2 = (a_0 - x_0) \tan \gamma_1$ follows:

$$x_0 = \frac{a_0 \tan \gamma_1}{\tan \gamma_1 + \tan \gamma_2} \qquad \text{and} \qquad c_0 = \frac{a_0 \tan \gamma_1 \tan \gamma_2}{\tan \gamma_1 + \tan \gamma_2}.$$

Similarly,

$$y_0 = \frac{a_0 \tan \delta_2}{\tan \delta_1 + \tan \delta_2} \qquad \text{and} \qquad d_0 = \frac{a_0 \tan \delta_1 \tan \delta_2}{\tan \delta_1 + \tan \delta_2}.$$

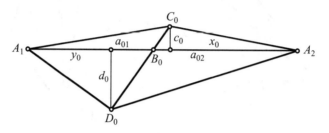

FIGURE 45
Ratio of parts

For the ratio a_{01}/a_{02} we next obtain

$$\frac{a_{01}}{a_{02}} = \frac{a_0 d_0 - x_0 d_0 + y_0 c_0}{a_0 c_0 + x_0 d_0 - y_0 c_0}.$$

If we substitute the values of x_0, c_0, y_0, and d_0 obtained above, we get

$$\frac{a_{01}}{a_{02}} = \frac{\tan \gamma_2 \tan \delta_2 (\tan \gamma_1 + \tan \delta_1)}{\tan \gamma_1 \tan \delta_1 (\tan \gamma_2 + \tan \delta_2)}.$$

Similarly,

$$\frac{a_{12}}{a_{10}} = \frac{\tan \gamma_0 \tan \delta_0 (\tan \gamma_2 + \tan \delta_2)}{\tan \gamma_2 \tan \delta_2 (\tan \gamma_0 + \tan \delta_0)}, \qquad \frac{a_{20}}{a_{21}} = \frac{\tan \gamma_1 \tan \delta_1 (\tan \gamma_0 + \tan \delta_0)}{\tan \gamma_0 \tan \delta_0 (\tan \gamma_1 + \tan \delta_1)}.$$

From this we deduce $\frac{a_{01}}{a_{02}} \frac{a_{12}}{a_{10}} \frac{a_{20}}{a_{21}} = 1$, and Ceva's Theorem implies the concurrence of the three straight lines $A_i B_i$.

Examples.

1. If we choose all six angles γ_i and δ_i, $i \in \{0, 1, 2\}$, to be equal, we obtain the centroid.

2. If, for $i \in \{0, 1, 2\}$, we choose $\gamma_i = \delta_i = \alpha_i$, then we obtain the intersection of the altitudes.

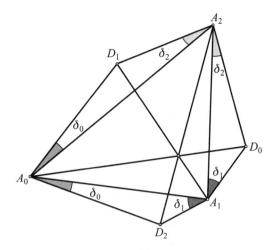

FIGURE 46
Jacobi's Theorem

3.5.2 Jacobi's Theorem as a special case

If we choose the special case $\gamma_i = \alpha_i$, $i \in \{0, 1, 2\}$, then $C_i = A_i$ and we obtain Jacobi's Theorem (Carl Friedrich Andreas Jacobi, 1795–1855) (Figure 46): *If, on the sides $A_{i+1}A_{i+2}$, $i \in \{0, 1, 2\}$, of a triangle $A_0 A_1 A_2$, triangles $D_i A_{i+1} A_{i+2}$ are constructed with angle $\angle D_{i-1} A_i A_{i+1} = \angle D_{i+1} A_i A_{i-1} = \delta_i$, then the three vertex transversals $A_i D_i$ are concurrent.*

3.5.3 Kiepert's Hyperbola

If, in Jacobi's Theorem, we choose the three angles δ_i to be equal, thus $\delta_0 = \delta_1 = \delta_2 = \delta$, there arises an intersection diagram with three similar isosceles triangles with their bases forming the triangle $A_0 A_1 A_2$ (Figure 47).

This idea leads to the Point of intersection 65.

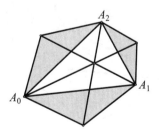

FIGURE 47
Superimposed similar isosceles triangles

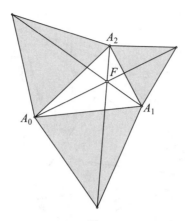

FIGURE **48**
The Fermat-point F

For $\delta = \frac{\pi}{3}$ we obtain the so-called Fermat-point F as point of intersection (Figure 48). The Fermat-point F has the property that (assuming none of the angles of the triangle is greater than $\frac{2\pi}{3}$) the network of paths formed by the segments $\overline{A_0 F}$, $\overline{A_1 F}$, $\overline{A_2 F}$, compared with any other network joining three vertices of the triangle, has the shortest total length (see [Coxeter 1963], p. 38 ff). Moreover, the three vertex transversals, in this case, intersect each other at equal angles of $\frac{2\pi}{3}$.

For $\delta \to 0$ we obtain the center of gravity and, for $\delta \to \frac{\pi}{2}$, the point of intersection of the altitudes. For $\delta \in [0, 2\pi]$ the point of intersection describes a hyperbola (Figure 49).

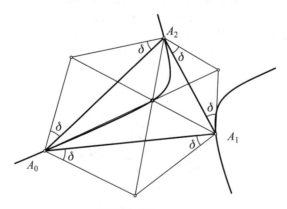

FIGURE **49**
Kiepert's hyperbola

This hyperbola is named for Wilhelm August Ludwig Kiepert (1846–1934) (see[Eddy/Fritsch 1994]). The hyperbola is rectangular and passes through the center of gravity, the point of intersection of the altitudes, and the Fermat-point, as well as through the vertices of the triangle. It contains the median of the three sides of the triangle as a chord on one branch of the hyperbola.

3.6 Remarks on selected points of intersection

In this section we offer commentaries, specimen proofs, and further remarks on selected points of intersection.

3.6.1 Point of intersection 32

This point of intersection is simply the point of intersection of angle-bisectors, since the angle-bisectors of a triangle also bisect the arcs of the circumcircle on the opposite sides. This follows from theorems on the angles at arcs of the circumference of a circle.

3.6.1.1 Points of intersection 36 to 40. We begin with three pairwise-tangent circles and place around them the "circumcircle" (Figure 50).

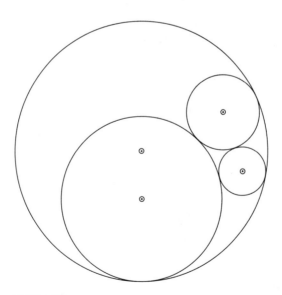

FIGURE **50**
Three pairwise-tangent circles and their "circumcircle"

In each of the circular triangles formed by two of the initial circles and the circumcircle we place an "incircle" (Figure 51).

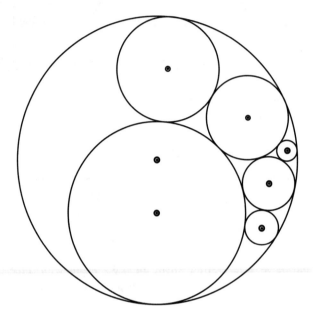

FIGURE 51
"Incircles" in the circular triangles

In this situation several points of intersection arise from various triples of straight lines (Figure 52). The reader is invited to find more.

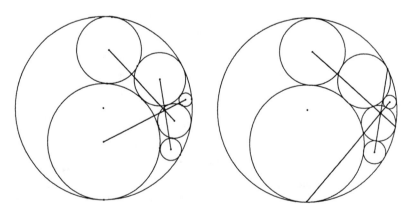

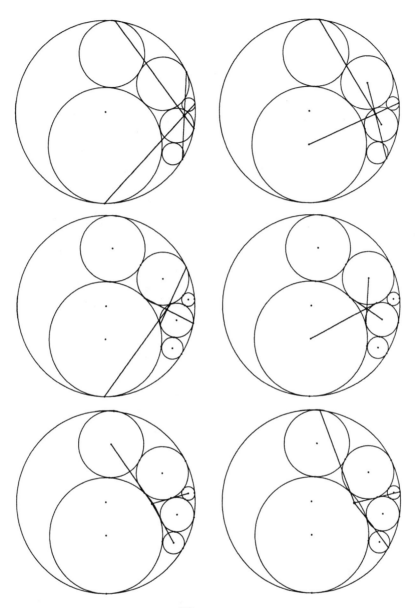

FIGURE 52
Various points of intersection

These eight points of intersection are collinear (Figure 53).

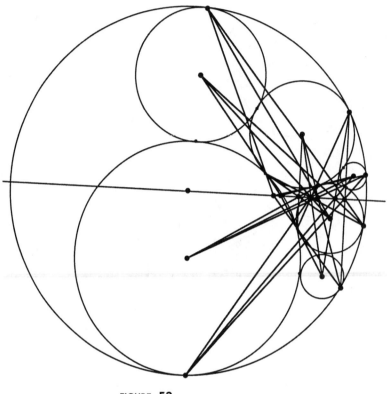

FIGURE 53
Collinear points of intersection

3.6.2 Point of intersection 79

It took the author several years to find a proof for this point of intersection. The proof he got is far from nice. It uses complex numbers, homogeneous coordinates, which are usually used in projective grometry, and a huge amount of calculation, executed with the help of a computer algebra system (Maple).

The author would be very happy if a reader found an elementary geometrical proof.

3.6.3 Point of intersection 84

3.6.3.1 Pythagoras' Theorem as a special case of the Law of Cosines.
A favorite theme of the independent work of my candidates for teaching
positions is to develop different proofs of Pythagoras' Theorem (see [Baptist
1997]), [Fraedrich 1995]). The usual proposal is to conclude the truth of
Pythagoras' Theorem from the special case $\gamma = 90°$ of the Law of Cosines.
A study of the familiar proofs of the Law of Cosines shows, however, that
almost all of these proofs use Pythagoras' Theorem. The resulting "proofs"
of Pythagoras' Theorem are merely vicious circles. Only the "vector" proof
of the Law of Cosines is "Pythagoras-free"; the study of vectors, however,
usually follows that of trigonometry in the curriculum.

3.6.3.2 A "Pythagoras-free" derivation of the Law of Cosines. On each
side of an acute-angled triangle ABC we draw, in the outward direction a
square (Figure 54). The altitudes of the triangle divide each square into two
rectangles.

The two rectangles denoted by I have the same area, namely[3], I $= bc \cos \alpha$.
Similarly II $= ca \cos \beta$, III $= ab \cos \gamma$.

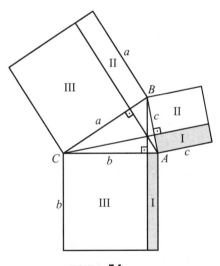

FIGURE 54
Attaching squares

---

[3] Translators' note: The author denotes by α, β, γ the angles of the triangle ABC.

Further

$$c^2 = \mathrm{I} + \mathrm{II} = (b^2 - \mathrm{III}) + (a^2 - \mathrm{III}) = a^2 + b^2 - 2\mathrm{III} = a^2 + b^2 - 2ab \cos \gamma.$$

3.6.3.3 Points of intersection.

We draw the perpendicular from the outer vertex A_1 (Figure 55) of the square onto the side BC of the triangle, and likewise the perpendicular from B_2 onto the side CA. Then the intersection of these two perpendiculars lies on the altitude h_c of the triangle. To prove this, we consider angles as in Figure 56.

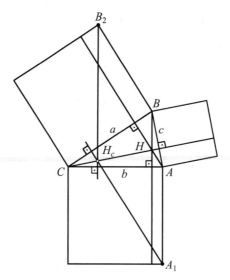

FIGURE 55
A point of intersection

Then on the one hand $\overline{F_a F_a^*} = b \sin \gamma$ and on the other hand $\overline{F_a F_a^*} = \overline{HH_c} \sin \beta$. From $b \sin \gamma = \overline{HH_c} \sin \beta$ and the Law of Sines it follows that $\overline{HH_c} = c$. The perpendicular from A_1 onto the side BC of the triangle therefore cuts the altitude h_c in a point at distance c from the point of intersection H of the altitudes. The corresponding reasoning for the perpendicular from B_2 onto the side CA of the triangle leads to the same distance calculation. Thus the three straight lines are concurrent.

Similarly we can drop perpendiculars from any of the external vertices of the squares and thus obtain six points of intersection, each of three straight lines (Figure 57). They form an affine-regular hexagon with the point of

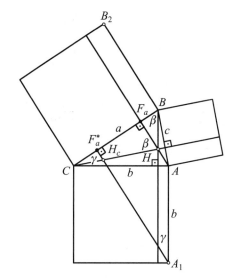

FIGURE 56
A proof figure

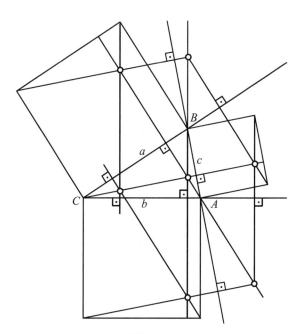

FIGURE 57
Six points of intersection

intersection of the altitudes as midpoint; this hexagon is put together from six triangles each congruent to the original triangle ABC.

Each one of these six points of intersection is an asymmetric point of intersection, in that the three concurrent straight lines do not all play the same role with respect to the original triangle. Each is a case of one altitude cut by two perpendiculars from outer vertices of a square.

3.6.4 Point of intersection 87

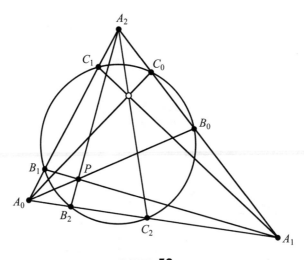

FIGURE 58
Proof diagram

The transversals $A_i B_i$, $i \in \{0, 1, 2\}$, are concurrent, so, by Ceva's Theorem

$$\frac{\overline{A_1 B_0}}{\overline{A_2 B_0}} \frac{\overline{A_2 B_1}}{\overline{A_0 B_1}} \frac{\overline{A_0 B_2}}{\overline{A_1 B_2}} = 1.$$

By the "power theorem" we have:

$$\overline{A_0 B_1}\, \overline{A_0 C_1} = \overline{A_0 B_2}\, \overline{A_0 C_2}.$$

Thus

$$\frac{\overline{A_0 B_2}}{\overline{A_0 B_1}} = \frac{\overline{A_0 C_1}}{\overline{A_0 C_2}}$$

and, analogously,

$$\frac{\overline{A_1 B_0}}{\overline{A_1 B_2}} = \frac{\overline{A_1 C_2}}{\overline{A_1 C_0}}, \quad \frac{\overline{A_2 B_1}}{\overline{A_2 B_0}} = \frac{\overline{A_2 C_0}}{\overline{A_2 C_1}}.$$

From

$$\frac{\overline{A_1 B_0}}{\overline{A_2 B_0}} \frac{\overline{A_2 B_1}}{\overline{A_0 B_1}} \frac{\overline{A_0 B_2}}{\overline{A_1 B_2}} = 1$$

it follows that

$$\frac{\overline{A_1 C_0}}{\overline{A_2 C_0}} \frac{\overline{A_2 C_1}}{\overline{A_0 C_1}} \frac{\overline{A_0 C_2}}{\overline{A_1 C_2}} = 1;$$

so that the transversals $A_i C_i$, $i \in \{0, 1, 2\}$, are also concurrent.

3.6.5   Points of intersection 96, 97, 98

Let two regular n-gons $A_0 A_1 \cdots A_{n-1}$ and $B_0 B_1 \cdots B_{n-1}$ have a common vertex, for example, let $A_0 = B_0$. Then the straight lines $A_i B_i$, $i \in \{1, 2, \ldots, n-1\}$ are concurrent (see Figure 59, with $n = 6$).

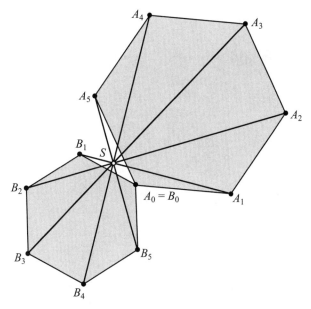

FIGURE 59
Two regular hexagons with a common vertex

Let S be the point of intersection of the straight lines $A_i B_i$ and $A_{n-1} B_{n-1}$. Then triangle $A_0 A_1 B_1$ is mapped onto the triangle $A_0 A_{n-1} B_{n-1}$ by a rotation around A_0 of $\frac{n-2}{n}\pi$. The two straight lines $A_1 B_1$ and $A_{n-1} B_{n-1}$ thus intersect in S at an angle of $\frac{n-2}{n}\pi$. The point S therefore lies on the arc on the segment $A_1 A_{n-1}$ subtending an angle of $\frac{n-2}{n}\pi$; this arc is however part of the circumcircle of the n-gon $A_0 A_1 \cdots A_{n-1}$. Similarly the point S also lies on the circumcircle of the n-gon $B_0 B_1 \cdots B_{n-1}$. The point S is thus, after A_0, the second point of intersection of the two circumcircles. In the same way one can show, using a rotation about A_0 of $\frac{n-2j}{n}\pi$, that the point of intersection of the two straight lines $A_j B_j$ and $A_{n-j} B_{n-j}$, $j \in \{1, 2, \ldots, \frac{n-1}{2}\}$, must coincide with this second point of intersection of the two circumcircles. For even n the straight line $A_{\frac{n}{2}} B_{\frac{n}{2}}$ must receive special attention, but this is the angle bisector of the two straight lines $A_{\frac{n}{2}-1} B_{\frac{n}{2}-1}$ and $A_{\frac{n}{2}+1} B_{\frac{n}{2}+1}$. Thus the asserted property of the point of intersection is proved. From the proof it follows that the straight lines $A_i B_i$ intersect at an angle which is an integer multiple of $\frac{\pi}{n}$. The straight line $S A_0$ also belongs to this regular sheaf of straight lines.

References

[Baptist 1992] Baptist, P., *Die Entwicklung der neueren Dreiecksgeometrie*, B.I. Wissenschaftsverlag, Mannheim, 1992.

[Baptist 1997] ——, *Pythagoras und kein Ende?*, Klett, Stuttgart, 1997.

[Berger 1987] Berger, M., *Geometry I*, Springer, New York, 1987.

[Buchmann 1975] Buchmann, G., *Nichteuklidische Elementargeometrie*, Orell Füssli, Zürich, 1975.

[Burg/Haf/Wille 1994] Burg, K., H. Haf, and F. Wille, *Höhere Mathematik für Ingenieure: Band IV Vektoranalysis und Funktionentheorie*, 2. Auflage, Teubner, Stuttgart, 1994.

[Butz 2003] Butz, T., *Fouriertransformationen für Fußgänger*, 3. Auflage, Teubner, Stuttgart, 2003.

[Cederberg 1995] Cederberg, J.N., *A Course in Modern Geometries,* Springer, New York, 1995.

[Chasles 1968] Chasles, M., *Geschichte der Geometrie,* Deutsche Übersetzung von L.A. Sohnke, 1839, Nachdruck, Wiesbaden, 1968.

[Coxeter 1963] Coxeter, H.S.M., *Unvergängliche Geometrie,* Birkhäuser, Basel, 1963.

[Detemple/Harold 1996] Detemple, D., and S. Harold, A Round-Up of Square Problems, *Mathematics Magazine,* Vol. 69, No. 1, February 1996, pp. 15–27.

[Donath 1976] Donath, E., *Die merkwürdigen Punkte und Linien des ebenen Dreiecks*, 3. Auflage, Deutscher Verlag der Wissenschaften, Berlin, 1976.

[Eddy/Fritsch 1994] Eddy, R.H., and R. Fritsch, The Conics of Ludwig Kiepert: A Comprehensive Lesson in the Geometry of the Triangle, *Mathematics Magazine*, Vol. 67, No. 3, June 1994, pp. 188–205.

[Filler 1993] Filler, A., *Euklidische und nichteuklidische Geometrie,* BI-Wissenschaftsverlag, Mannheim, 1993.

[Fraedrich 1995] Fraedrich, A. M., *Die Satzgruppe des Pythagoras,* BI-Wissenschaftsverlag, Mannheim, 1993.

[Hartshorne 2000] Hartshorne, R., *Geometry: Euclid and Beyond,* Springer, New York, 2000.

[Hauptmann 1995] Hauptmann, W., Erzeugung "merkwürdiger Punkte," *PM Praxis der Mathematik* 37, 1995, S. 8.

[Heuser 1992] Heuser, H., *Funktionalanalysis*, 3. Auflage, Teubner, Stuttgart, 1992.

[Hoehn 2001] Hoehn, L., Extriangles and Excevians, *Mathematics Magazine*, Vol. 74, No. 5, December 2001, pp. 384–388.

[Holme 2002] Holme, A., *Geometry: Our Cultural Heritage*, Springer, New York, 2002.

[Jänich 2001] Jänich, K., *Mathematik I: Geschrieben für Physiker*, Springer, Berlin, 2001.

[Kimberling 1998] Kimberling, C., Triangle Centers and Central Triangles, *Congr. Numer.* 129 (1998), pp. 1–295.

[Kinsey/Moore 2002] Kinsey, L.C., and T.E. Moore, *Symmetry, Shape and Space: An Introduction to Mathematics Through Geometry*, Key College Publishing, Emeryville, 2002.

[Klemenz 2003] Klemenz, H., Merkwürdiges im Dreieck, *VSMP Bulletin, herausgegeben vom Verein Schweizerischer Mathematik- und Physiklehrer*, No. 91, Februar 2003, S. 16–23.

[Kroll 1990] Kroll, W., Rundwege und Kreuzfahrten, *PM Praxis der Mathematik* 32, 1990, S. 1–9.

[Lenz 1967] Lenz, H., *Nichteuklidische Geometrie*, Bibliographisches Institut, Mannheim, 1967.

[Longuet-Higgins 2001] Longuet-Higgins, M.S., On the principal centers of a triangle, *Elemente der Mathematik* 56, 2001, S. 122–129.

[Madelung 1964] Madelung, E., *Die mathematischen Hilfsmittel des Physikers*, Springer, Berlin, 1964.

[mathforum] mathforum.org/dynamic.html

[mathworld] mathworld.wolfram.com/topics/TriangleCenters.html

[Meyberg/Vachenauer 2001] Meyberg, K., and P. Vachenauer, *Höhere Mathematik I*, 6. Auflage, Springer, Berlin, 2001.

[Nöbeling 1976] Nöbeling, G., *Einführung in die nichteuklidischen Geometrien der Ebene*, Walter de Gruyter, Berlin, 1976.

[Schumann 1990/91] Schumann, H., Geometrie im Zug-Modus: Teil 1, *Didaktik der Mathematik* 18, 1990, S. 290–303. Teil 2, *Didaktik der Mathematik* 19, 1991, S. 50–78.

[Teubner-Stiftung] www.stiftung-teubner-leipzig.de/1-mathematik.htm

[Walser 1990–1994] Walser, H., Schlußpunkt, *Didaktik der Mathematik* 18 (1990) bis 22 (1994), jeweils letzte Heftseite.

[Walser 1991] ——, Ein Schnittpunktsatz, *Praxis der Mathematik* 33, 1991, 70–71.

[Walser 1993] ——, Die Eulersche Gerade als Ort "merkwürdiger Punkte," *Didaktik der Mathematik* 21, 1993, S. 95–9.

[Walser 1994] ——, Eine Verallgemeinerung der Winkelhalbierenden, *Didaktik der Mathematik* 22, 1994, S. 50–56.

[Walser 1998] ——, *Symmetrie*, Teubner, Stuttgart, 1998.

[Walser 2000] ——, *Symmetry*, Translated by Peter Hilton and Jean Pedersen, Washington, DC, MAA, 2000.

[Walser 2001] ——, *The Golden Section*, Translated by Peter Hilton and Jean Pedersen, Washington, DC, MAA, 2001.

[Walser 2003] ——, Eine Schar von Schnittpunkten im Dreieck, *Praxis der Mathematik* 2/45, 2003, S. 66–68.

[Walser 2004] ——, *Der Goldene Schnitt,* EAGLE 001, 4. Auflage, EAGLE-EINBLICKE, Edition am Gutenbergplatz Leipzig, Leipzig, 2004.

[Wells 1991] Wells, D., *The Penguin Dictionary of Curious and Interesting Geometry*, Penguin Books, London, 1991.

[Zeitler 1970] Zeitler, H., *Hyperbolische Geometrie*, Bayerischer Schulbuch Verlag, München, 1970.

Index

About the Author

Hans Walser was born in 1944 in Rheineck, Switzerland. He received his PhD in 1975 from the ETH Zurich (Swiss Federal Institute of Technology, Zurich). From 1975 to 2001 he was a teacher at the State College of Thurgau, Switzerland, and he was a visiting guest at Santa Clara University, California, in 1996 and 1998. Walser is now a lecturer at the ETH Zurich and the University of Basel. His primary research interest is geometry. He is a member of the Swiss Mathematical Society, the MAA, the AMS, and several German societies devoted to mathematics and teaching of mathematics. Walser is the author of numerous books. The MAA has already published translations of two other books: *The Golden Section* and *Symmetry*.